Electrical
Installation
Work Level 1

Peter Roberts

Routledge
Taylor & Francis Group

LONDON AND NEW YORK

First published 2018
by Routledge
2 Park Square, Milton Park, Abingdon, Oxon OX14 4RN

and by Routledge
711 Third Avenue, New York, NY 10017

Routledge is an imprint of the Taylor & Francis Group, an informa business

British Library Cataloguing-in-Publication Data
A catalogue record for this book is available from the British Library

Library of Congress Cataloging-in-Publication Data
Names: Roberts, Peter, 1963 March 13- author.Title: Electrical installation work. Level 1 / Peter Roberts.Description: EAL edition. I Abingdon, Oxon ; New York, NY : Routledge, 2017. I Includes bibliographical references and index.Identifiers: LCCN 2016059206I ISBN 9781138232068 (pbk. : alk. paper) I ISBN 9781315303550 (ebook : alk. paper)Subjects: LCSH: Electric apparatus and appliances--Installation--Textbooks. I Electric wiring, Interior--Textbooks.Classification: LCC TK452 .R5749 2017 I DDC 621.319/24--dc23LC record available at https://lccn.loc.gov/2016059206

ISGN: 978-1-138-30978-4 (hbk)
ISBN: 978-1-138-23206-8 (pbk)
ISBN: 978-1-3153-0355-0 (ebk)

Typeset in Helvetica Neue
by Servis Filmsetting Ltd, Stockport, Cheshire

Printed and bound by CPI Group (UK) Ltd, Croydon, CR0 4YY

Electrical Installation Work

Level 1

The best up-to-date textbook for EAL's Level 1 Diploma in Electrical Installation (601/0409/0)

- Fully up to date with the 3rd Amendment of the 17th Edition IET Wiring Regulations
- Expert advice ensured to cover what learners need to know in order to pass their exams or complete their assignments
- Extensive online material to help both learners and lecturers

Written specifically for the EAL Diploma in Electrical Installation, this book has a chapter dedicated to each unit of the syllabus. Every learning outcome from the syllabus is covered in highlighted sections, and there is a checklist at the end of each chapter to ensure that each objective has been achieved before moving on to the next section. End of chapter revision questions will help you check your understanding and consolidate the key concepts learned in each chapter. Fully up to date with the 3rd Amendment of the 17th Edition Wiring Regulations, this book is a must have for any learner working towards EAL electrical installation qualifications, also providing an insight to those who are considering a career in the electrical installation or construction industry.

Peter Roberts is an ex-RAF Chief Technician and is currently an electrical installation lecturer, as well as an EAL question writer, based in Coleg Menai, Bangor, North Wales.

ACKNOWLEDGEMENTS

I am extremely grateful to some of the editorial and production staff at Routledge involved in creating this publication, namely: Tony Moore, Seth Townley and Scott Oakley. Equally, I would like to thank Paul Partington and Pete Olito, for fulfilling their promise to proof read, despite their own work commitments. It would also be remiss of me to forget the continual support of my family: my son Garym who came to the rescue with several professional photographs and my daughter Anwen who took time off from her own studies to review many aspects. Special mention must go to my long suffering wife Sian; without her understanding and encouragement, completing such a project would not have been possible.

A Big thank you to you all!

Contents

Health and safety in electrical installation

EAL Unit ELEC1/01

EAL Electrical Installation Work – Level 1. 978-1-138-23206-8
© 2017 P. Roberts. Published by Taylor & Francis. All rights reserved.

Learning outcomes

The learner will:

1. Know the fundamental aspects of health and safety legislation that apply to electrical installation.
2. Know how to recognise and respond to hazardous situations.
3. Know the basic safe working procedures.
4. Know how to respond to accidents that occur while working.
5. Know the basic procedures for electrical safety.
6. Know the methods of safely using access equipment.
7. Carry out basic safe working practices.
8. Know fire safety procedures.

Learning outcome 1.1

Outline the aims of general health and safety legislation in protecting the workforce and members of the public.

Learning outcome 1.2

Recognise the responsibilities of persons under health and safety legislation.

It is virtually impossible to work in industry today without having to sit through several induction packages or courses designed to prepare an individual for the workplace. A work induction programme, however, should never be thought of as a chore, because this process has taken over 200 years to evolve and even today strives to keep all work-related environments as safe as possible.

During the 1800s, although investigations into workplace accidents were carried out, unfortunately this was not because employers were necessarily interested in their employee's welfare; their involvement would have been directed towards addressing why productivity had been reduced. Although at this time ready replacements could soon be found from what were called workhouses, replacing injured workers was still seen as an inconvenience – the substitute employees needed training, which meant that it took time for productivity to return to normal.

Early laws were finally introduced with regard to the working conditions of child labour in the textile industry through the Factory Act 1802. They limited workhouse children from working at night and capped their working day to 12 hours. This was eventually extended to all workers under the age of 18 through the Factories Act of 1831.

A significant development came about through a further Factories Act in 1833, when factory inspectors were appointed and tasked to overlook how child textile workers were treated and ensure that employers were behaving responsibly. These factory inspectors were given powers, which are similar to the Health and Safety Inspectors of today, since they were permitted to enter mills and actually question workers. They were also able to influence the need to provide guards on machinery as well as recording accidents. It is important to remember that new manufacturing processes were part of the Industrial Revolution, which created great wealth for the government and certain employers. However, it also included widespread abuse of those who were employed within it because the workers of that time had very few human rights. In 1837, it was finally established in law that an employer owes a duty of care to employees and this remains true to this day.

Legislation is passed through Parliament so that it becomes law, in order to enforce its message as well as prosecuting those who do not comply. The Health and Safety at Work Act 1974 (HASWA) was a key piece of legislation, known as an enabling act. This allows the government to pass further legislation known as statutory regulations, which are tied to the HASWA without the need to pass them through Parliament.

The HASWA 1974 also reinforces that responsibilities regarding safety apply to both employers and employees and acts as a reminder that each person has a duty of care to each other and to the public. This is very important, since ignorance of the law is no excuse – if you fail to carry out your duty and injure

Important point

Legislation is passed through Parliament to become statutory law to protect employees.

Tackle a difficult word: enabling

Allows further regulations to be passed

Important point

Statutory law exists to protect employees.

another person, or even yourself, then the law will punish you. This means that everyone working on a site must recognise their responsibilities when it comes to health and safety. A major statutory regulation that stemmed from the HASWA was the Electricity at Work Regulation 1989; its purpose was to ensure that employers put measures in place to protect their employees from being subjected to electrocution. Measures such as:

- Ensuring safe isolation procedures are in place
- Ensuring that employees are adequately supervised
- Ensuring that all test equipment is approved and appropriate

It is important to distinguish between statutory regulations such as the Electricity at Work Regulations 1989 and BS 7671 the wiring regulations. The wiring regulations are not legal documents; they are non-statutory but contain technical information that is written in such a way as to comply with any legal aspect. A prime example is that all electrical circuits should be provided with fault protection in order that a protective device will operate and isolate the circuit. However, the actual details regarding how this should be carried out are very technical, and are not contained in the Electricity at Work Regulations 1989, but lie within BS 7671. By always referring to the Wiring Regulations, an electrician is following what is known as their code of practice.

Responsibilities

The HASWA is written in such a way that an employer will bear more responsibility than an employee because they must ensure that the workplace is safe as far as is reasonably practicable. In other words, they must take all measures possible to protect their workforce and this includes:

- Providing personal protection equipment (PPE) that is appropriate to work activities
- Providing appropriate training irrespective of the costs involved
- Providing any specialised equipment and tools required for a task
- Continually assess and review workplace hazards

A key objective for any employer to meet is to install safe systems of work; in other words, every effort should be made to ensure that the potential of harm from hazards is either removed or reduced.

Once these safe systems of work are in place, then the employee has a duty of care to abide by them and under no circumstance interfere with them even if this means the job takes longer. For instance, if the employer has identified that PPE such as a safety harness is to be worn whilst working at height, then the employee must wear it. Remember that failure to carry out safety instructions can not only injure yourself and other employees, but possibly harm members of the public or other important visitors being escorted around a site.

Case study

Let us look at a case study involving an employer and a car owner who, after experiencing problems, drove her car to a local garage to be fixed. Having explained the problem, she left the car in the hands of the mechanic who also happened to be the garage owner. The garage owner took charge of the car, drove it up and on to a ramp and then raised it up to examine it underneath. The

Did you know?

The definition of a professional person is someone who is paid to do a job of work.

Figure 1.1 Both employers and employees are responsible for embracing health and safety procedures

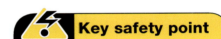

Key safety point

By law, an employer must provide all relevant equipment required for the workplace; the employee however must wear or use it.

Did you know?

You can be prosecuted for injuring yourself.

Key safety point

An employee must co-operate with their employer in maintaining safety including voicing their concern if any safety issues become apparent in the workplace

Key safety point

Employers must carry out risk assessments and include everyone that could be affected, even customers.

customer, however, had failed to tell the mechanic that there were children asleep in the back of the car. Unfortunately, one of the children woke up, left their seat, opened the car door and fell to the ground – sustaining serious facial injuries.

Question: Who was at fault?

It might surprise you to learn that when applying Health and Safety law, the garage owner was at fault, since from the moment he was handed the keys and took possession of the car, he had a legal requirement to carry out a risk assessment. Had he done so, he would have noticed the children asleep in the back. Why not look at other employers that have been prosecuted by clicking on the following link: Prosecution – HSE RSS feeds: http://news.hse.gov.uk/category/about-hse/prosecution/page/30/

This was a serious accident so an investigation was undertaken by an independent organisation; known as the Health and Safety Executive (HSE), it can be thought of as the health and safety police. However, unlike the police they do not need a warrant to enter premises or interview employees.

The HSE stated that the garage owner had failed in his duty of care regarding the customer and her children.

Key safety point

An employee must never interfere with safety systems.

In later years, further European workplace-related regulations, also known as statutory instruments, were introduced and are referred to as the Six Pack Regulations. The HASWA is therefore sometimes referred to as an umbrella act, with the Six Pack Regulations sitting underneath it. Figure 1.2 lists the Acts contained within the Six Pack Regulations.

- Health and Safety (Display Screen Equipment) Regulations 1992
- Personal Protective Equipment at Work Regulations 1992
- Management of Health and Safety at Work Regulations 1992
- Workplace (Health, Safety and Welfare) Regulations 1992
- Manual Handling Operations 1992
- Provision and Use of Work Equipment Regulations 1992

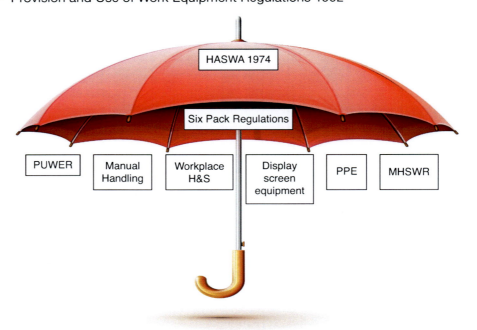

Figure 1.2 The Health & Safety Act 1974, is sometimes referred to as an umbrella act

We will now consider each in turn.

1. **Health and Safety (Display Screen Equipment) Regulations 1992**

This regulation came about in order to safeguard and specify the minimum safe working conditions required for people who work with display screen equipment such as computers. An employer must carry out a workspace assessment to ensure that all of the following equipment is appropriate for use:

- Display screen
- Keyboard
- User space
- Chair
- Lighting (glare)
- Software

Improper glare and continual exposure to a computer screen can cause headaches, which is why the employers have a duty of care to pay for regular eye checks. Improper chairs, especially when sitting for long periods can cause serious back pain, therefore the chairs used must be appropriate for this kind of work, in order to support a person's posture when seated.

Top tip

Health and Safety is all about applying common sense.

2. **Personal Protective Equipment Regulations 2002**

The Regulations place a duty of care on the employer to provide appropriate PPE with regard to the work activity but it must also be 'fit for purpose'. In the UK, PPE should also meet other requirements such as being stamped with a CE mark, which proves that it complies with products sold in Europe.

3. **Management of Health and Safety at Work Regulations (MHSWR) 1999**

This regulation makes sure that employers/managers:

- Carry out risk assessments to highlight workplace hazards
- Write technical procedures to indicate how certain jobs should be carried out
- Employ special permits when employees are required to work in hazardous environments
- Employ specialists if the company or its employees are not qualified in any particular area
- Nominate responsible people to specialise and promote good practice in various areas and disciplines
- Provide information and training to employees
- Co-operate on health and safety matters with other employers who share the same workplace
- Talk with their employees on a regular basis
- Review all their policies and procedures on a regular basis

4. **Workplace (Health, Safety and Welfare) Regulations 1992**

This regulation ensures that all places of work are safe, by ensuring that all facilities, equipment and systems are maintained in good working order.

Key safety point

Employers must provide safe systems of work.

5. **Manual Handling Operations Regulations 1992**

By law, employers must:

- Carry out an assessment of all operations which involve lifting
- Provide adequate training with regard to promoting correct lifting techniques
- Provide mechanical lifting equipment when required

- Ensure that drivers and operators of elevated work platform such as cherry pickers are trained, authorised and current

6. Provision and Use of Work Equipment Regulations 1998 (PUWER)

PUWER exists to make sure that the tools and equipment in use are appropriate by making sure that:

- They are suitable to the task they are involved in
- They are maintained to stay fit for purpose
- Employees are given training if necessary
- Safety devices or procedures including warning signs are in place

Question?

Which of the Six Pack Regulations did the garage owner fail to comply with regarding the customer and her child? There are **TWO** answers

The answer

- Management of Health and Safety at Work Regulations 1992
- Workplace Health, Safety and Welfare Regulations 1992

Remember that the Management of Health and Safety spells out that safe systems of work must be in place and these will include procedures such as risk assessments in order to identify any hazards and especially adopting procedures to protect customers. All workplaces should be organised in such a way that both pedestrians and vehicles can operate in a safe manner.

Time out to discuss what we have learnt about the Health and Safety at Work Act 1974:

- It is a statutory Act designed to protect the workforce
- Employers must provide:
  - PPE, training, suitable and appropriate equipment as well as safe systems of work
- Health and Safety has to be policed by the HSE, in order to ensure that both employers and employees meet their:
  - Duty of care to themselves and others

To review your understanding try this True or False quiz based on certain requirements of health and safety.

Question 1: The Health and Safety Act 1974 is **not** a legal document.

True	False

Question 2: Certain work-based regulations are known as the Six Pack Regulations.

True	False

Question 3: Employees should, by law, have to provide their own PPE.

True	False

Learning outcome 2.1

List the types of general workplace hazards that may be encountered while at work.

Learning outcome 2.2

Recognise the methods that can be used to prevent accidents or dangerous situations occurring during work activities.

The previous learning outcomes discussed how Acts of Parliament and statutory regulations have evolved to protect the workforce in a working environment. Despite this, work related accidents still happen. Statistically the most common cause of accidents are caused through slips, trips and falls; whilst the main cause of deaths in the construction industry stems from working at height.

An employer should ensure that they apply a rigorous system of assessing, preventing and controlling their workplaces from contributing or causing accidents. An employer must ask themselves:

- Can any procedure be changed so that it involves less risk?
- Can any known hazardous substance be replaced with a safer alternative?
- Are the guards or barriers currently used adequate?
- If a hazard cannot be removed, what PPE must be provided to reduce the risk from injury?

All of these elements go towards answering a basic question: is there anything more that can be done to make the job safer?

Let us explore a working environment and at each stage think about what might go wrong, because even simple activities may be hazardous. Certain basic working practices can reduce the number of accidents and their impact.

Husbandry

Husbandry is a vital working practice since people tend to be hoarders. For instance, in certain circumstances, as materials left in waste paper baskets decompose, it can cause problems. Some have even been known to spontaneously combust. Waste, therefore, should be disposed of in a proper manner including separating out hazardous waste, recyclable waste and waste that cannot be recycled which should be sent to landfill.

Figure 1.3 Proper husbandry keeps the workplace not only clean and tidy but safe

Description

Husbandry: A clean and tidy workplace is a safe workplace.

That said, any unused materials should be returned to stores; not only will this keep down an employer's running costs, it does not impact so negatively on the environment.

Tools and equipment should be checked before and after use to make sure they are clean, complete and still fit for purpose. Equipment should have a permanent storage location, which will stop it becoming a trip hazard or, if left unattended, possibly blocking emergency exits. Workstations, benches and surrounding floor areas should be free of dirt or swarf (waste pieces of metal, wood, or plastic) through ensuring that a daily cleaning programme is performed at the end of the day, unless it is necessary before that.

Trailing leads can cause an immediate trip hazard, as well as possibly incurring a risk of electric shock, which means that they should be routed away from any walkways, preferably installing them along existing fence lines.

Slippery or uneven surfaces can lessen surface grip both when walking or when operating mechanical lifting devices. Any spillage involving lubricants such as petrol, diesel or other substances such as oils are a major cause of slip accidents. The same applies to uneven surfaces since they offer less friction and can also cause serious injuries.

All floors should, therefore, be regularly cleaned and warning signs should be erected in order to alert personnel of any uneven surfaces. Some workplaces even have a spillage plan to ensure that when large amounts of fluid such as petrol, oils and lubricants are involved they are dealt with through a proper process without incurring further injury or incident.

Dust and fumes present a hazard regarding inhalation (breathing in) which means that appropriate PPE must be issued in any circumstance when it is not possible to remove such a hazard. Do not forget that if the employer highlights and provides PPE then the employee must cooperate and wear it. Dust-rich environments, if exposed to an ignition source, could potentially cause an explosion, therefore it will be necessary to position appropriate fire-fighting equipment in certain locations.

Depending on the substance, when handling contaminants and irritants ill effects can range from mild irritation of the skin to severe pain and loss of limbs and even loss of life. COSHH risk assessments need to be deployed in order to highlight any specific hazards which could be encountered during its use, as well as highlighting when control measures such as PPE are required.

Handling and transporting equipment or materials

Most cases of back pain are not caused by disease but by minor sprains, strains or injuries, brought about through incorrect manual handling techniques. A very high percentage of injuries are back related, therefore all manual handling activity needs to be reviewed to see if mechanical lifting devices are required, as well as promoting good lifting techniques when lifting lighter loads. All storage areas need to be kept tidy so that they do not become trip hazards in themselves and if materials are required to be stored at height then control measures such as guardrails need to be in place.

The main cause of deaths in the construction industry occurs through activities involving working at height and largely involve employees using unsupported or incorrect access equipment. This means that any activity above ground must be scrutinised through a full risk assessment process in order to pinpoint, for instance, if safety harnesses are required. The operation of scissor lifts or cherry

Top tip

Tools and equipment should be checked before and after use.

Description

COSHH: Control of Substances Hazardous to Health.

Figure 1.4 A construction site is potentially very dangerous, which is why it's known as a Special Location

pickers is controlled so that only authorised and trained personnel are permitted to operate them.

Hazardous malfunctions of equipment

If a piece of equipment fails in its function (that is, fails to do what it is supposed to do) and, as a result of this failure, has the potential to cause harm, then this would be defined as a hazardous malfunction.

All work equipment must be:

- Suitable and safe for its intended use
- Maintained in a safe condition
- Used only by instructed persons
- Provided with suitable safety measures, protective devices and warning signs
- Used correctly and appropriately and given safe storage when not in use

When equipment does become defective, for whatever reason, then it needs to be removed from use, by being labelled unserviceable and physically removed from service, being electrically isolated or segregated by barriers.

That said, if any employee is asked to carry out a procedure and thinks it too dangerous to continue, then they should raise the alarm by reporting the matter immediately to their supervisor; for example, if they notice that machine guards have been removed by an employer in order to speed up production.

Another good working practice is to use tool caddies as shown in Figure 1.5, so that all tools in use are controlled. It can also act as a temporary bin to store waste until such time as it can be disposed of correctly.

The use of shadow boards, an example of which is shown in Figure 1.6, can help identify and ensure that all tools are returned after use. This reduces the

Key safety point

Employees should always refer any safety issues to their immediate supervisor.

Figure 1.5 Tool cadies keep your tools under control

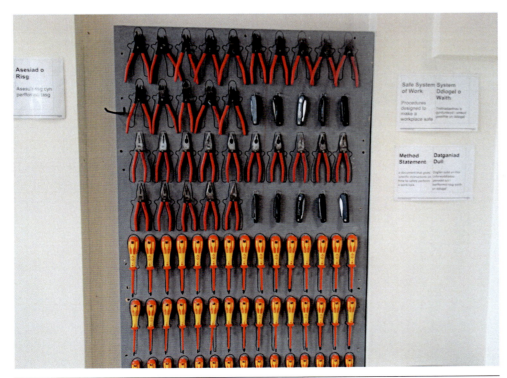

Figure 1.6 Shadow boards indicate when a tool is missing or lost

possibility of a tool becoming a trip hazard or even injuring another employee, as well as possibly coming into contact with machinery or live electrical supplies and causing further damage.

Risk assessments have already been mentioned as a means of bringing about a safe system of work, but we will now expand on how this is carried out alongside two other key safety procedures.

Risk assessments

This is a procedure that tries to identify which employees are subjected to work-based hazards and introduce control measures to reduce the possibility of them being harmed. The HSE recommends a 5-step approach:

- Identify any hazards
- Decide who might be harmed and how
- Evaluate the risks by comparing current control measures against precautions
- Record any significant findings
- Review your assessment and update if necessary

Method statement

A method statement is a written procedure detailing how a work activity is going to be carried out, including identifying any special tools or equipment required.

Permit to work

A permit to work is a procedure that is used in particularly hazardous environments. Before any work can be carried out, the work must be authorised by the permit and cleared after the task has taken place.

Key safety point

The definition of a hazard is anything with a potential to cause harm.

Description

A method statement is a systematic procedure.

Description

A permit to work is a procedure used in hazardous environments.

General health and safety activity

This activity contains two sets of matching terms and each matching pair should be given the same number. The first one has been done for you: 'The Health and Safety Act 1974 = 1' matched with 'Enabling act that acts as a health and safety umbrella = 1'

The Health and Safety Act 1974	Manual Handling Regulations 1992	Enabling act that acts as a health and safety umbrella	Regulation that ensures employers train their employees with correct lifting techniques
= 1	= 2	= 1	=
Personal Protection Equipment Regulations 2002	Workplace Health and Safety Regulations 1992	Regulation that ensures employers provide PPE	Regulation that ensures employers provide a safe working environment
=	=	= 3	= 4

Signs

Risks and hazards can also be highlighted in the workplace through various workplace signs and fall into five different categories. They include:

1. Blue Mandatory signs can be thought of as 'You Must'. For example, Figures 1.7 and 1.8 indicate that you must wear protective footwear and eye protection, respectively.

Figure 1.7 Mandatory signs can be thought of as Must DO, such as you MUST wear safety footwear

Figure 1.8 Remember eye protection must be appropriate to a particular task

2. Warning signs on the other hand, highlight specific hazards. For example, Figure 1.9 emphasises that dangerous 'High Voltage' supplies are located nearby.

Figure 1.9 Warning signs indicate that potentially lethal hazards are located nearby

? Did you know?

Warning signs are also used in nature such as the South American frog shown in Figure 1.10, which is deliberately vibrant in order to warn other animals that it is poisonous. A warning sign is therefore equally brightly coloured in order to raise awareness of a particular hazard.

3. Prohibition signs such as that shown in Figure 1.11 can be thought of as 'You Must Not; for example in the example shown: 'You must not swim here'.

Figure 1.10 Even nature uses Yellow and Black to highlight potential danger

Figure 1.11 Some signs are bilingual, but the colours Yellow and Black always indicates potential hazards and Red literally signposts danger

Figure 1.12 The colour Green is associated with a Safe Condition sign

4. A Safe Condition sign contains important safety information such as indicating which route should be taken to exit a building during an emergency or highlighting where the nominated assembly point is. An example is shown in Figure 1.12.

5. A fire sign is used to indicate where certain fire-fighting equipment or alarm points are located. They are always coloured red as exhibited by Figure 1.13.

All these different measures help identify and deal with general workplace hazards and recognise the methods that can be used to prevent accidents or when dangerous situations occur during work activities.

Figure 1.13 Red is used to indicate the positioning and location of Fire Fighting Equipment

Safety sign activity

This activity contains two sets of matching terms and each matching pair should be given the same number. The first one has been done for you: 'Mandatory sign = 1' matched with 'Blue 'must do' = 1'

Mandatory sign	Prohibition Sign	Blue 'must do'	Red and white 'must not'
= 1	= 2	= 1	=
Warning sign	Safe condition sign	Yellow and black 'be aware of hazardous'	Green 'safety information'
=	=	= 3	= 4

Learning outcome 2.3

Recognise warning symbols of hazardous substances.

Learning outcome 2.4

Identify the general precautions necessary for working with commonly encountered substances in building engineering service.

It depends on the substance, but different illnesses can come about through contact with what are classified as contaminants and irritants and these can range from mild irritation of the skin to severe pain, loss of limbs and, in severe cases, loss of life. Hazards are defined as things with the potential to cause harm and anybody that is exposed to a hazard is at risk.

COSHH stands for Control of Substances Hazardous to Health and is a type of risk assessment designed specifically for when someone is handling and using substances. When substances are manufactured, the manufacturer will produce what is called a data sheet, in which the manufacturer will indicate and highlight any particular risks involved such as being potentially harmful through ingestion (eating), absorption (entered through skin), inhalation (through breathing) or even that the substance is potentially explosive. The data sheet will also indicate, where necessary, if any control measures must be in place, such as insisting that the substance should only be used in locations that contain proper ventilation or an extraction system. Appropriate PPE will also be listed, including indicating that it should be made from suitable materials that can safeguard the user during its use. For example, when handling petrol, oils and lubricants any PPE should be resistant or impervious to their corrosive effects.

Occasionally wearing gloves is difficult when carrying out manual tasks, but a person's skin can become sensitive to oils and other substances, which can cause dermatitis. During such circumstances, it is advisable that the person use barrier cream, applied directly on to the hands, as it will protect the skin from infection.

Key safety point

A COSHH procedure is a type of risk assessment involving substances.

Tackle a difficult word: dermatitis

Skin condition caused when the skin reacts to a substance

Explosive

Flammable

Oxidizing

Compressed Gas

Corrosive

Toxic

Irritant

Environmentally
Damaging

Health Hazard

Figure 1.14 CLP symbols are recognised worldwide

Apart from COSHH risk assessments, a workplace will also use signs or pictograms to indicate the presence of hazardous chemicals. The pictograms shown above (Figure 1.14) were introduced in 2015, and replace previous versions so that the UK would come into line with globally recognised symbols known as CLP. This stands for Classification, Labelling and Packaging of substances.

Learning outcome 2.5

State what action to take should a hazardous situation occur while at work.

Earlier we mentioned that dangerous situations must be immediately reported to a supervisor. However, if your supervisor is not around or not contactable then any person can raise the alarm and warn other people.

Let us look at an example.

Question?

Having noticed an oil spillage on the floor do you:

Response 1: Rush off and find a mop and bucket and clear up the oil before anybody is injured?

Response 2: Raise the alarm so that others in the area are alerted to the danger and ask them to find a mop and bucket while you stand guard?

Both responses are admirable but only Response 2 is correct. The reason for this is that if you leave the area unguarded any person walking through will not be aware of the slip hazard and could injure themselves.

Top tip

If you witness a potentially dangerous situation always, report it to your immediate supervisor.

Case study

A similar situation happened to me whilst climbing into a Hawk aircraft to rectify a reported electrical system fault. However, just as I was about to enter into the cockpit, I noticed that one of the ejection seat safety pins was not inserted correctly and was effectively hanging off. This meant that the seat firing handle could be accidently pulled, which would mean that certain explosive cartridges would operate and the seat along with anybody positioned in the cockpit would be ejected into the hanger roof. I was not trained or authorised to reinsert the safety pin and, unfortunately, the people I required were located in another building. I therefore chose to stay and guard the aircraft, stopping anybody else from entering the cockpit, and called out for someone else to raise the alarm and summon help.

Remember, if you ever find yourself in a tricky situation talk to your supervisor or ask someone else for help!

Time out to consider some important facts and key points:

- All workplaces will include some hazards
- Hazards are defined as something with the potential to cause harm
- Employers should instigate safe systems of work, which will include carrying out risk assessments, which will need to be reviewed to ensure if various hazards originally identified still exist or have changed
- Employees require training on certain equipment
- Husbandry includes checking tools and equipment before and after use, cleaning them if necessary and storing them in their correct locations
- Disposing of waste is vital, especially adhering to procedures regarding hazardous waste

Lastly, various signs will:
- Indicate the presence of hazards (warning signs)
- Reinforce that certain practices must be met (mandatory signs) or must not be carried out (prohibition signs)
- Highlight any safety information, especially with regard to an emergency situation (safe condition sign)

To review your understanding try this True or False quiz based on Health and Safety systems.

Question 1: Wearing gloves or using barrier cream can reduce dermatitis.

True	False

Question 2: A mandatory sign should be thought of as 'must do'.

True	False

Question 3: COSHH is a system that controls the use of substances.

True	False

Learning outcome 2.6

Identify the situations where asbestos may be commonly found in the workplace.

According to the HSE, asbestos has been the main cause of occupational ill health from about 1950 onwards and past practices are now responsible for about 4,000 people dying from asbestos-related cancers every year.

Asbestos is a very good insulator and products were made from three types: 'blue asbestos' (crocidolite), 'brown asbestos' (amosite) and 'white asbestos' (chrysotile). All three are dangerous, but blue and brown asbestos are known to be more hazardous than white. Although now banned, all three were used extensively in different products and materials such as:

- Fire breaks in ceiling voids and ducts
- Thermal insulation of pipes
- Ceiling panels and tiles
- Paper and paper products used for insulation of electrical equipment
- Cement products
- Early forms of artex
- Corrugated sheets used in roofing and wall cladding
- Floor tiles

Asbestos is safe if left undisturbed, but drilling or cutting into it causes small particles to be released into the air and it is these particles, which people breathe in, which can lead to asbestosis and eventually lung cancer. The human body cannot get rid of any inhaled asbestos particles; therefore, any new exposure simply builds on the previous exposure. This is known as a cumulative effect.

There is no cure for asbestosis

The HSE recommends that if you have doubts about any material that you come in contact with, it is best to presume that it contains asbestos. It is therefore vital that you do not break or damage the material, but simply remove yourself from the area and report the matter to your supervisor.

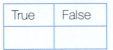

Key safety point

Asbestos becomes dangerous if it is disturbed.

Tackle a difficult word: cumulative

Increasing, growing, snowballing

The Control of Asbestos Regulations 2012 require:

- Mandatory training for anyone liable to be exposed to asbestos fibres at work
- Samples should be taken when it is necessary to identify if any material contains asbestos, but only by suitably trained people
- Asbestos should only be removed by licensed removal contractors (Figure 1.15)
- Employers should carry out risk assessments to ascertain the condition of any asbestos as well as using notices and stickers to highlight the possibility of the presence of asbestos

Case study

Review the life of Avril Grant who lost two husbands to asbestos, one of whom was an electrician: www.hse.gov.uk/asbestos/casestudies/avril-grant.htm

Learning outcome 3.1

State the purpose and application of protective equipment.

PPE exists to limit the extent of an accident but it cannot stop accidents from happening. PPE is therefore used as last resort when it is simply not possible to remove a particular hazard and all other options of keeping employees safe have been considered.

Employers not only have a responsibility to provide PPE but should ensure that PPE is being worn, including using mandatory signs as a reminder. In order to conform to Personal Protective Equipment Regulations 1992/2002, PPE should include a CE mark, which means it complies with the use of equipment in Europe. When issued, employees have a responsibility to store it safely and check it regularly for damage and/or wear.

PPE is normally associated with parts of the body that need protection such as:

Eyes (when drilling, chiselling masonry surfaces or when using machinery)

Combat the hazard by wearing:

- Safety glasses
- Face shield

Foot protection (when there is a risk from falling objects or hazardous substances)

Combat the hazard by wearing:

- Steel toe caps
- Safety shoes impervious to oil

Leg protection (exposure to hazardous substances or severe weather)

Combat the hazard by wearing:

- Leather apron if handling batteries
- Gaiters – weather/water protection

Hands and arms (when there is a risk to the hands from sharp objects or hot/cold surfaces)

? Did you know?

According to the HSE, on average 20 tradespeople die of asbestos related illnesses each week. This means four plumbers, six electricians and eight joiners.

Figure 1.15 Asbestos must be removed by licensed and specialist contractors

? Did you know?

PPE is only required when the hazard cannot be removed.

? Did you know?

The employer should provide PPE but the employee should not only wear it but also maintain it in good condition and store it away when not in use.

? Did you know?

PPE should be appropriate to the task.

Combat the hazard by wearing:

* Gloves, gauntlets (Figure 1.16)
* Barrier cream – dermatitis
* Coveralls

Head (when there is a risk from falling objects)

Combat the hazard by wearing:

* Helmet
* Neck guard

Figure 1.16 Some electricians' gloves can withstand voltages as large as 36,000 volts

Lungs (when there is a risk from ingestion of particles including dust)

Combat the hazard by wearing:

* Respirators
* Face masks
* Forced air systems

Hearing protection from excessive noise

Combat the hazard by using:

* Ear defenders
* Ear plugs

Body protection required

Combat the hazard by using:

* High visibility vests (when hazards exist from moving machinery and plant)
* Coveralls (overalls)
* Wet weather gear

Remember an employer must provide PPE whenever there is no possibility of removing anything that has the potential to cause harm. Let us look at a case study, which contains significant employer and employee failings.

Case study

Ken Woodward

Ken worked in a fizzy drinks factory and was tasked with cleaning out some appliances. This process was normally carried out using a pre-mixed solution, but unfortunately the company had run out several weeks before. Instead, employees had reverted to mixing two separate solutions in open containers: one a cleaning agent and the other a type of bleach.

Ken was asked to do this job at the end of his night shift so that the day shift could go straight into production. However, Ken had never attempted this procedure before and had not received any training. He also had not been made aware of at least two unreported near misses when other people had been injured attempting the same procedure. Neither was there any PPE readily to hand for anyone to attempt this procedure.

The two solutions were perfectly safe if you mixed them in a certain order, but became unstable if the solutions were blended in the opposite way. Because there was no training involved, Ken was unaware of any danger and an explosion occurred so violent that the acid in the solution burned into his eyes and subsequently blinded him.

This is a classic case of why 'some of the best people make the worst kinds of mistake'. Ken's life was saved by two workmates: one grabbed him and shoved him into an industrial shower, trying desperately to wash the solution away from Ken's eyes; the other was a maintenance engineer who, being dedicated in his work duties and staying late at the end of his last shift, had repaired a broken industrial shower. Once the shower was fixed, it needed testing and this meant stepping directly into the shower to ensure it operated automatically, getting wet in the process. This was the shower used to save Ken's life. Thankfully, the maintenance man firmly believed that all employees have a duty of care to each other and when working in industry you need to follow something through.

Questions that need to be asked about this case study:

The company had run out of the normal mixing solution several weeks before and it had **NOT** been re-ordered – why?

Employees were allowed to make their own cleaning solution without any training or PPE – why?

Two near misses had gone unreported – why?

The answer to these questions is that people do sometimes adopt a 'can do' attitude, and whilst this is commendable, it's vital that they consider if they can do it safely. Employers have a duty of care to look at their procedures, especially when working with hazardous substances. Sometimes it seems that all that matters is meeting production targets, but there is a cost involved when procedures are either not followed correctly or, as in this case, were not even put in place. This was a life-changing injury which, in turn, affected Ken's family. It also affected some of his workmates who were involved on the day, some of whom still have difficulty coming to terms with what happened.

Ken now goes around various industries and openly talks about the use of PPE but also encourages others to look at and challenge people's behaviour regarding safety in the workplace, so that no one else suffers in the same way as he did. He has helped a great deal of people, mostly because people tend to listen to any safety message given by Ken. For instance, after visiting one particular factory, one industrial worker decided to wear his safety goggles for the first time, despite having carried out a particular activity a hundred times before without incident. On this occasion however, a battery on a forklift truck exploded, which is why he wrote to Ken to thank him for saving his sight. This is a perfect example of a learning culture.

Ken's moving story can be found at http://kenwoodward.co.uk/kens-accident/

Personal protection equipment activity

This activity contains two sets of matching terms and each matching pair should be given the same number. The first one has been done for you: 'Hard hat = 1' matched with 'Head protection = 1'

Hard hat = 1	Hand protection =	Head protection = 1	Gloves = 2
Ear defenders = 3	Safety glasses = 4	Steel toe caps = 5	Eye protection =
Safety shoes =	Other than gloves this can protect against dermatitis = 6	Barrier cream =	Hearing protection =
Hi-viz vest =	Protect your whole body = 8	Be seen, be safe = 7	Overalls =

Learning outcome 3.2

State the procedures for manually handling heavy and bulky items.

Manual handling is the technical term for moving a load which is an object having a mass of 20 kilogrammes or more. Moving such loads is not always straightforward. For instance, they may:

- have an awkward shape
- be slippery or have sharp edges

This means that loads may be difficult to grip. If a task is assessed as being appropriate to handle manually, then proper lifting techniques must be used, especially since it is known that lifting and moving heavy equipment is one of the biggest single causes of personal injury.

Never make do with a situation – if the load is too heavy, get assistance or use mechanical lifting gear to carry out the task. Remember, employers have a duty of care to ensure safe systems of work are employed and this includes providing appropriate equipment or machinery, including training in its use.

When moving loads:

1) Assess the load – look for any sharp edges
2) Decide if protective clothing or safety footwear is required

3) Plan the route and ensure that it is free from obstacles and obstructions
4) Assess if the load will pass through a door
5) Use correct manual handling techniques such as keeping your back straight, knees bent, head and chin up
6) Ensure both feet including your heels are firmly placed on the floor, one foot slightly in front of the other

If the load is heavy, you should also apply 7 and 8:

7) Assess if a mechanical lifting device such as a sack trolley (Figure 1.17) is required
8) Ensure load is kept stable and secured

When you are assessing any type of load against mechanical lifting gear, you must ensure that the equipment's Safe Working Load (SWL) is not exceeded.

In addition to the points listed above, it is very important that a supervisor is appointed for the task, someone who stands back and oversees the operation. When loads are being moved through corridors, then it is equally important that a safety person known as a Look Out is put in place, so that they can warn any passer-by of the approaching hazard. When very large loads are being delivered, especially if being moved by machinery such as a crane, then barriers should be used to keep people away from harm.

Key safety point

Always check Safe Working Load.

Learning outcome 4.1

Identify the actions that should be taken when an accident or emergency is discovered.

When an emergency happens, it is all very well to say, 'Do not panic', but in truth people do.

It is actually hardwired into our brains to respond in two different ways to an emergency, often called fight or flight. Therefore, depending on our reaction, we might tackle the situation head on or run away. Neither of these is an ideal approach, though; it is far better to think fast but act slow, so that you give yourself time to think of a plan of action. One of the reasons why new employees are given an induction process is so that they are then equipped to deal with certain emergencies. For instance, an induction programme will state:

* Where all the emergency stop buttons are located so that the electrical supply can be isolated
* Where the first aid points are located as well naming all first aid qualified personnel
* Where fire alarm panels and fire-fighting equipment are located, so that individuals are able to raise the alarm and also tackle any fire, as long as they do not put themselves in danger
* Where designated escape routes lie as well as nominated assembly points

Some companies actually employ a fire response hat system, whereby certain persons in authority wear different coloured hats. For instance, an individual wearing a Green hat would be the nominated person responsible for coordinating the evacuation of employees. Various section heads would therefore report to this person in order to establish if all employees are accounted for. A Red hat would designate the person responsible for taking command of a firefighting or first aid team.

Figure 1.17 A risk assessment will often highlight if loads should be lifted by using mechanical lifting gear such as a sack trolley

Key safety point

After an evacuation, always carry out a roll call at the assembly point to ensure that all employees are present.

It is important to remember that contacting the emergency services should never be attempted within the building that is on fire. This is because smoke travels faster than fire and the person making the phone call could easily be overcome.

Having made contact from a safe location, always give as much detail and information as possible, such as:

- Which building the fire is in, including a number or name if possible
- The type of fire
- The number of casualties
- Any suspected missing employees

The fire service, therefore, know in advance what lies ahead, including the possibility that they might need to enter the building.

Sounding the alarm is not always straightforward either, since some alarm systems are local, which means that they are not necessarily linked to the emergency services. Therefore, it is always a good idea to shout 'fire, fire, fire' when operating any emergency switching, so that you can alert others. You also need to be made aware that certain establishments, such as military camps, coordinate their telephone exchanges; therefore, despite dialling locally you might find yourself speaking to someone in Scotland. In all situations, common sense should be used.

Important point

Always give the emergency services as much relevant information as possible.

Learning outcome 4.2

Outline the procedures for dealing with minor/major injuries that can occur while working.

Despite all the training and safety precautions that are introduced in industry, accidents happen on a daily basis. You cannot qualify as a first aider by reading a book, but being aware of certain methods and techniques might enable you to assist a casualty or a nominated first aider at some point in the future. For example, being aware of how a statement such as 'never assume, check' can apply to an accident scene. There have been numerous cases where a large number of people have attended a road traffic accident but, unfortunately, everyone has assumed that someone else has called the emergency services.

You can also reduce the effect of shock by reassuring a casualty that help is on the way. This simple act can possibly save someone's life.

First aid

On finding an unconscious casualty such as that shown in Figure 1.18, the acronym HABC can be used as an aid to how to apply first aid.

H is for Hazard

You must assess the area around the casualty, just in case the hazard is still present. When suspecting electrocution such as that shown in Figure 1.18, your first action must be to locate the source of danger and look to safely isolate the supply without injuring yourself. This should be done by operating emergency isolation switches or, failing that, separate the casualty from any live cable by using a wooden object such as a brush. It is important to stress that by all means try and help a casualty, but never put yourself at risk.

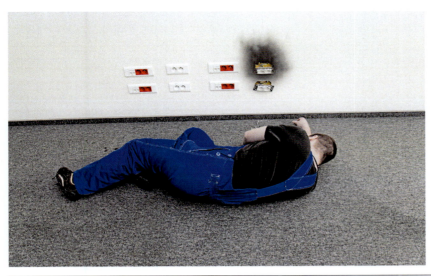

Figure 1.18 If you suspect that a person has been electrocuted, then try to switch the supply off, but only if you can do so without endangering yourself

A is for Airway

Establishing an airway is vital and this can be done by both supporting and extending their head in order to ensure that they have not swallowed their tongue or other objects such as food stuff or even false teeth.

B is for Breathing

Place your ear next to the casualty's mouth in order to sense if they are breathing. You can also look down at their chest to see if it is rising. Blue lips are a symptom of a lack of oxygen in the body (cyanosis).

C is for Circulation

Measure the casualty's pulse, which will establish if their heart is working. Use your fingers when measuring a pulse, but not your thumb. This is because a thumb contains its own pulse and you can easily think that the casualty has a healthy pulse rate.

If you cannot establish that they have a pulse, then you will need to carry out the following CPR method:

* Lay the casualty on their back and kneel down beside them.
* A good tip for finding the heart is to follow the ribs up to the top of the rib cage and then place two fingers above it. Place the heel of one hand over the target for compression.
* Interlock your fingers and straighten your arms then press down to about 4–5 cm for an adult (Figure 1.19).
* You should press down hard and fast at a rate of one push per second. Alternatively, and dependant on your age, use the rhythm of *Staying Alive* by the Bee Gees ('Ha, Ha, Ha, Ha, staying alive, staying alive'). It can be seen on YouTube as demonstrated by Vinnie Jones in a TV advertisement for the British Heart Foundation. Younger readers might be better suited to use the chorus of 'Nelly the elephant'.
* After 30 repetitions check if the casualty has regained a pulse. If not give, two rescue breaths.
* Maintain their airway by supporting their chin upwards, close off their nose and after having taken a breath, and place your mouth over the casualty's mouth endeavouring to maintain a seal.

Key safety point

Always use the HABC response to first aid.

Important point

CPR should be applied using the ratio of 30 compressions to two rescue breaths.

- Breathe steadily for approximately one second, looking down their chest wall to see if it is rising. If the chest is not rising, this indicates a possible blockage.
- Continue the sequence of 30 chest compressions followed by two rescue breaths.
- If the casualty recovers or regains a pulse, stop the heart massage.

Look for any other injuries such as bleeding:

- Treat the casualty for shock
- Place them in the recovery position

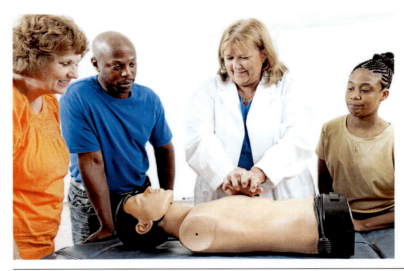

Figure 1.19 CPR should be applied - 30 compressions to 2 rescue breaths

Recovery position

With casualty lying on their back

Lift their right arm and place it by the casualty's left cheek

Keep arm in position

Lift their right leg up by their knee and roll the casualty forward

Ensure their airway is maintained

Monitor breathing and pulse continuously

If injuries allow, turn the casualty to the other side after 30 minutes

Cuts

Minor cuts and abrasions can be treated with plasters or dressings. Remember to inform the responsible person so that any first aid supplies can be replenished.

Minor burns

Burns caused by heat or chemicals should be treated by applying cold running water over the wound for at least 10 minutes. Although clingfilm can be applied, ensure that any dressing is not fluffy; otherwise the casualty will experience further pain when it has to be removed and cleaned.

Eye injuries

When a foreign object enters someone's eye, then eyewash bottles should be used to wash out the eye affected. Eyewash bottles are normally located either alongside first aid kits or above industrial sinks.

Important point

Always put an unconscious casualty in the recovery position.

Bleeding

When treating a bleed then the acronym PEEP can be used:

P: Position yourself next to the casualty

E: Examine the wound

E: Elevate the wound

P: Pressure should be applied through a dressing
If the casualty continues to bleed apply a further dressing, as seen in Figure 1.20.
Do not remove the first one.

Figure 1.20 PEEP: Position yourself, Examine the wound, Elevate the limb, apply Pressure

Treatment for shock

- Lay the casualty down, raise, and support their legs
- Keep them warm
- Do not give them anything to eat or drink
- Talk to them reassure them that help is on the way
- Check their breathing and pulse frequently

Time out to consider some important facts and key points:

- Talking and reassuring a casualty can reduce the effects of shock
- Asbestos is safe unless you disturb it
- If in doubt treat any material in the same way as asbestos
- PPE is only required when you cannot remove a hazard
- PPE must be appropriate to the task; there is no one size fits all situation

Important point

If the casualty continues to bleed, keep applying further dressings.

Important point

Talking to a casualty and reassuring them can reduce the effect of shock.

- Correct manual handling involves using the correct lifting technique for a particular load
- Always walk in an emergency and ensure you are fully aware of evacuation procedures
- Moving heavy loads should be carried out by using mechanical lifting equipment
- When applying first aid, use HABC (Hazard, Airway, Breathing and Circulation)
- Never assume that someone else has called the emergency services – check

To review your understanding, try this True or False quiz based on first aid and safety procedures.

Question 1: Your first action when attending to a casualty is to ensure you are not in danger.

True	False

Question 2: It is better to remove a hazard than wear PPE.

True	False

Question 3: Burns caused by heat or chemicals should be treated by applying cold running water for at least 10 minutes.

True	False

Question 4: As an employee, you must be aware of designated escape routes as well as nominated assembly points.

True	False

Question 5: Asbestos is safe as long as you do not get it wet.

True	False

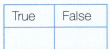

Important point

Accidents are recorded so that people can learn from past mistakes.

Accident reporting

All accidents, however minor, have to be both reported and recorded. Companies tend not to use accident books these days but record the detail on accident forms. Failing to record a minor accident that results in the wound being infected at a later date could stop any further compensation claims in the future.

Near misses

Although a near miss is basically an incident and not an accident, employees are encouraged to record the details surrounding the event. This means other employees can read the reports to learn from other people's mistakes and stop incidents from becoming accidents.

This is also why a dangerous incident, which did not cause harm but had the potential to cause severe injury, is reported. Encouraging employees to report such incidents helps bring about a learning culture.

Reporting dangerous occurrences is actually a statutory requirement known as RIDDOR, which stands for Reporting of Injuries, Diseases and Dangerous Occurrences Regulations. This means that employers must raise, record and report any serious workplace-related accidents.

Important point

Near misses are recorded so that incidents do not become accidents.

Learning outcome 5.1

Identify the common electrical dangers encountered on construction sites and in dwellings.

When a person gets an electric shock, they basically become part of the circuit. Electricity can be thought of as ET (Extra-terrestrial, Figure 1.21) because like ET electricity wants to find its way home – back to the supply transformer.

Unfortunately, a human body is made from approximately 60% water and given that water is an excellent conductor of electricity this means the person completes the circuit. The extent of an electric shock will depend on the value of the current and length of the exposure but it is possible that as little as 50 mA (0.05 amp) can kill a person. There are many ways to receive an electric shock and we will consider these next.

Faulty equipment

Electrical equipment can malfunction and therefore become a serious threat to the user. If any metal part of an electrical equipment becomes live then we rely on an earthing system to isolate the supply.

An earthing system is purposely designed to be very low in resistance, in order to offer a fault current an easy route to ground. In other words, electricity always

Figure 1.21 Electricity always wants to find its way home

Important point

When someone is electrocuted, the person becomes part of the circuit as shown in Figure 1.22.

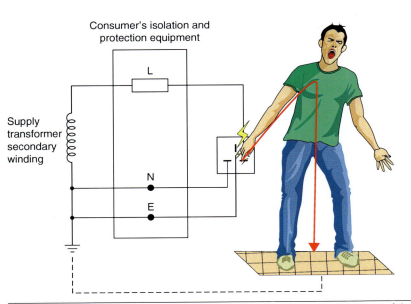

Figure 1.22 When a person receives an electric shock, they become part of the circuit

travels down the path of least resistance and will flow down an earthing system rather than through a person. When this occurs, the value of the fault current is very high and a circuit protective device such as a fuse or circuit breaker operates in order to isolate the supply. However, if the earthing system fails, then a fault current would have no other option but to flow through a person that has had the misfortune of coming into contact with it.

Overhead cables

Overhead cables tend not be insulated and in fact use air as an insulator. Any person coming into contact with overhead cables whilst standing on the ground would be electrocuted instantly. Scaffolding and metallic ladders would obviously conduct, but electrocution has also occurred through fishing rods because they are often made from carbon fibre. Carbon is a good conductor of electricity.

Electricity is also distributed underground, which means that it is possible for a person to make contact with these types of supplies when excavating the ground.

There are other ways of protecting users, such as class 2 equipment which encases all the live parts in two sets of barriers and insulation so that it should not be possible for anyone to touch any live supply. Because electrocution is therefore not possible, class 2 equipment does not require an earthing system.

Another method of reducing the effect of an electric shock is to use what is known as a reduced voltage system. Reduced voltage systems ensure that all hand-held electrical equipment on a construction site is powered through 110 V and not through 230 V. These type of connectors are referred to as commando sockets, which are designed to be robust and splash proof. All the voltage systems are colour coded, with 110 V being represented by yellow as shown in Figure 1.23.

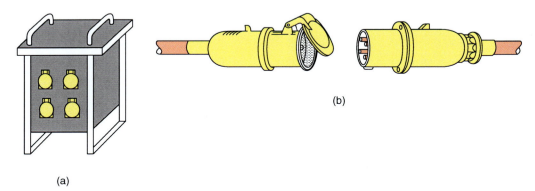

(b)

(a)

Figure 1.23 All hand held equipment on a construction site must be supplied through a yellow 110 V industrial plug

Description

SELV stands for Separated Extra Low Voltage.

When certain areas, however, contain a lot of moisture or dampness, then equipment should be supplied through a Separated Extra Low Voltage (SELV) system. A SELV system is very effective because it operates on voltages less than 50 V in value as well as supplying electricity through a safety isolation transformer.

A further safety device, known as a Residual Current Device (RCD), operates if it senses that current is travelling down the earthing system, rather than being returned to the supply through the neutral conductor as it would normally. Various ratings of RCD are installed in different locations. For instance, socket outlets must be protected by an RCD rated at 30 mA, but other ratings such as 100 mA are also used when safeguarding the input supplies to 110 V safety transformers.

Please note that the safest electrical system in use would be battery-operated equipment, since they operate on 12 V d.c.

We have covered many ways through which a person can receive a potentially lethal electric shock but it can also occur through damaged or worn electrical cables. A cable is normally made of three separate parts: the conductor, which carries the current; the inner insulation which isolates all the conductors from each other; and the outer sheath, which protects the cable from any mechanical impact. Damage to any of the elements will seriously weaken it but also potentially kill anybody that came into contact with it. There may, however, be certain signs of damage such as that seen in Figure 1.24.

Figure 1.24 Overheating can lead to electrical fires

There is also a need for all electrical equipment used on site to be periodically inspected. This process is known as PAT – which stands for Portable Appliance Testing – and will be carried out by a qualified responsible person. Prior to use, however, the user should always carry out a pre-use check to ensure that the equipment is physically safe to use.

A pre-use check should include ensuring that:

* The equipment has a current portable appliance test label attached to it. An example of which is shown in Figure 1.25
* There are no bare wires showing
* The cable's insulation is not damaged. Note that it is especially important to ensure that bare conductors are not exposed
* The plug top is in good condition including an appropriately rated fuse
* The grip cord is clamped down on the outer sheath and not on the inner insulation
* The outer case is not damaged
* There are no missing screws
* There are no signs of overheating

A very handy rhyme to help you remember this sequence is shown below. Check the:

* Case
* Face
* Cables
* Labels
* Extras (accessories)

Figure 1.25 Labels can prove that a piece of electrical equipment have been electrically tested or highlight that it is Not to be used

Safe isolation procedure

Most deaths from electrocutions occur because an electrical supply that had been made dead was made live, despite the fact that there was a person working on the system. The HSE devised GS 38, a guide to enable electricians to bring about best practice when installing electrical systems. This document insists that only approved test equipment should be used to check that a circuit is dead and that certain popular devices that are easily available to purchase such as volt sticks, neon fitted screwdrivers and even ordinary multi-meters are not authorised for such tasks. The reason for this is that most of these devices will certainly indicate that a circuit is live, but are not necessarily sensitive enough to recognise if a circuit is operating in a low voltage state. Approved test equipment, however, is designed to be sensitive enough to recognise all levels of voltage and also incorporates built in safety features such as finger guards and spring loaded tips to stop a person accidently making contact with live supplies. An example of an approved voltage indicator is shown in Figure 1.26.

A safe isolation procedure also uses other important equipment as shown in Figure 1.27. A proving unit is used to check if the tester is actually working before

Figure 1.26 Only Approved Test equipment is permitted to ensure that an electrical circuit has been made dead

Figure 1.27 Electrical circuits need to be locked off to ensure that the electrical supply is not accidently switched back on

it is used to check if the supply is dead. However, it must also be used one final time to ensure that it has not failed during the proving process. In other words, you have to test the tester twice. Various locking off devices are used to ensure that circuit breakers and fuses are locked off safely and, alongside an appropriate safety sign, will ensure that the supply cannot accidently be reapplied.

Time out, let's summarise some **key** points!

- Tools form a very important part of electrical work, therefore it is good practice to check tools and equipment before and after every use. This will allow time to replace any faulty or damaged items
- All accidents have to be reported, even a near miss when nobody was harmed
- Reporting accidents and incidents helps people to learn from the mistakes of others
- To receive an electric shock you must become part of the circuit
- Reduced voltage systems help reduce the effect of an electric shock

Learning outcome 6.1

Outline the situations where it may be necessary to work at height.

New Work at Height Regulations came into force on 6 April 2005 and they specify that any work above ground level is classified as working at height. Given that the main cause of deaths in the construction industry stems from this kind of activity, every precaution should be taken to ensure that the access equipment is appropriate and fit for purpose.

There is a range of access equipment in existence. It is important to understand:

- When it should be used
- How to use it safely
- How to look after it
- How to adopt any relevant safety precautions

Ladders and stepladders are both considered within the same classification in that they are only acceptable for work over short durations, such as for tasks no more than 30 minutes long. Before use, however, always carry out a risk assessment in looking for any obvious hazards such as soft or uneven ground and especially overhead cables.

It is also vital that you inspect the actual equipment before use. The following pre-use check applies to ladders.

Check that:

- Stiles are secure
- Rungs are secure
- Feet rubbers are in place
- Wooden ladders are not painted (hidden faults)

Important point

Ladders should be used for light work only approximately 30 minutes in length.

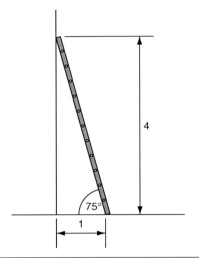

Figure 1.28 Ladders should be erected to an angle of 75 degrees

Key safety point

When using ladders, always maintain three points of contact.

Figure 1.29 A trestle scaffold is very useful if the work surface happens to be uneven

If a double ladder is being used, ensure:

- Guide and latching brackets are secure

Ladder use

Do:

- Use for 'light work' only, for a maximum of 30 minutes
- Always maintain three points of contact (hands and feet) at the working position
- Erect ladders at 75° (Figure 1.28)
- Secure ladders in place (tied)
- Footing a ladder is a last resort
- Take care with overhead lines – carry out a risk assessment
- Keep your body position central within both ladder styles

Do not:

- Overload (carry too much weight); consider SWL
- Overreach (move the ladder to a new position)
- Over run 30 minutes

Stepladders

Most of the rules that apply to ladders apply equally to stepladders, especially in carrying out a risk assessment and pre-use checks to establish that it is in good working order. A stepladder does include extra elements such as ensuring that the cord is undamaged. If the stepladder is fitted with a hinge mechanism then ensure it is secure and extends and locks in place correctly.

Trestle scaffold

A trestle scaffold is shown in Figure 1.29 and contains a pair of trestles spanned by scaffolding boards. Two boards at least 450 mm wide must be used to make the platform and it is advisable if the platform is more than 2 m above the ground then toe boards and guardrails must be fitted. Trestle scaffolds are very effective when you encounter uneven ground since they can be adjusted to match the gradient.

Mobile scaffold towers

Although mobile scaffold towers can be constructed from basic scaffold it is more commonly assembled from light alloy tubes. A typical arrangement is shown in Figure 1.30. The tower is built up in sections by first connecting and fixing an intermediate platform and then extending up to build a second tower if required.

If its height extends above 2 m then for safety purposes this type of platform must be fitted with toe boards and guardrails to act as a safety measure, whilst also reducing the chance of objects or tools being kicked off and injuring any persons passing by. When in use, it is advisable to lock the wheels but when moving the platform it should be pushed at the base. A ladder is normally used to gain access to the platform, but it should be tied for security and extended 1 m or five rungs above the working platform.

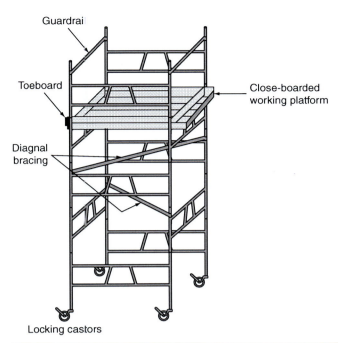

Figure 1.30 Toe boards and guardrails lessen the chance of tools and materials falling off and injuring anybody walking past

Mobile elevated work platforms

Aerial work platforms (AWP) are types of mechanical device used to gain access to equipment situated at high locations. A scissor lift is one type of AWP and would be ideal to replace a light fitting that was located centrally in the middle of a warehouse ceiling.

Another AWP is commonly called a cherry picker such as that shown in Figure 1.31. All AWP could potentially encounter overhead power lines, but appropriate precautions, such as carrying out a risk assessment, are particularly

 **Key safety point**

Cherry pickers should only be operated by trained and authorised personnel. They are particularly vulnerable in high winds and when overhead cables are located nearby.

Figure 1.31 Only trained and authorised workers are permitted to operate a cherry picker

relevant to a cherry picker given the height that its bucket can extend to. Further consideration include wind strength and should not be used during these conditions for fear of the vehicle becoming unsteady and unstable. The actual person in the basket should also wear and use a harness, which should be attached to the safety cage.

Like many other machines, only authorised and trained personnel are allowed to operate it and furthermore approved operators have to be frequently re-authorized, usually every six months.

Roof ladders

These type of ladders are used when carrying out installation work on sloping roofs. Many have wheels fitted which means securing is only a matter of hooking them over the roof ridge. They are also referred to as cat ladders.

Crawling boards

Working in attic spaces is potentially hazardous for many reasons. During hot summer days, the heat can be stifling, made worse if the house has been insulated with thermal insulation. Access to the attic space is not guaranteed unless a proper roof ladder or stairs have been installed and even when you climb up, the area might be deficient of lighting. All these potential problems make working in attics dangerous, but especially if the floor has not been boarded over. A word of warning, it is never a good idea to try and negotiate your way by stepping over roof joists. One slip could cause a serious injury, severe inconvenience to the householder, as well as a lot of embarrassment to the employer. The employer might have liability insurance but their subscription will increase following any subsequent claim. Crawling boards therefore can be used to ensure that you are secure underfoot.

Scaffolding is seen as the ultimate access equipment when supporting a long-term project, but although it can be tied to a building, it must be self-supporting. Only approved contractors are allowed to erect or dismantle scaffolding and it must be checked every seven days.

Learning outcome 8.1

Outline how combustion occurs.

Learning outcome 8.2

State the method for tackling small localised fires that can occur in the workplace.

Learning outcome 6.7

Define how combustion occurs.

Important point

Removing any single element required to create or maintain a fire will extinguish it.

For any fire to start it needs three different components, which means that if one of these components is removed the fire will be put out. All three basic components are shown by the use of a fire triangle (Figure 1.32).

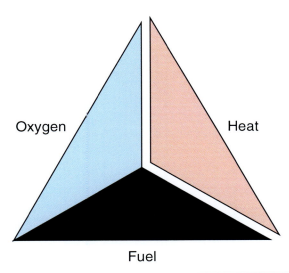

Figure 1.32 A fire cannot survive if it's starved of oxygen, deprived of fuel or its heat is cooled

Fuel

Fuel can be thought of as liquids such as petrol, oils and lubricants. It can also extend to gas as well as any combustible solid material such as wood, cardboard or paper. In other words, anything that can burn.

Because of this, flammable materials should be stored in cool, secure storage and only accessed by authorised employees.

It is also one of the reasons why husbandry is so important – such as the cleaning up of waste material and disposing of it in an appropriate manner – as this reduces the possibility of a fire spreading. Effectively this means we are depriving it of fuel.

Oxygen

Oxygen is all around us because it is a major component in the air that we breathe, but can be eliminated from a small fire by smothering it with a fire blanket, sand or even foam. This is why a foam fire extinguisher is so effective with a fuel fire, since it spreads and forms a carpet of foam across its surface and cuts off its oxygen supply.

Limiting oxygen supply is also why all doors and windows should be shut during a fire evacuation, but only if you can do so without putting yourself in danger. It is also why fire doors are strategically placed, in order to seal off specific areas and slow down the build-up of smoke and heat. It goes without saying, therefore, that fire doors should never be left open or ajar.

Heat

Heat can be removed by cooling the heat being generated, normally through a water extinguisher. It must be stressed that water will make a situation worse if used on certain types of fires, especially those involving electricity and liquids such as petrol or oil.

Chip pan fires used to be commonplace and there have been reported cases of water being thrown over the chip pan, which only served to scatter the burning hot oil around the kitchen and actually start new fires. To tackle a chip pan fire carry out the following action:

- Turn the heat off
- Use a damp cloth (cloth that has been put under water and wringed through) and lay over the chip fan
- Do not move the chip pan

Fires are divided into six classes or categories, each matched with an appropriate fire extinguisher as indicated in Figure 1.33.

- Class A fires involve solid materials such as wood, paper and fabrics
- Class B fires involve liquids such as petrol, oils and lubricants
- Class C fires involve gas or liquefied gas
- Class D fires involve burning metal (not a common classification)
- Class E fires involve faulty electrical equipment
- Class F fires involve fat and cooking oil

It has already been stated that you should only tackle a fire if you can so without putting yourself at risk and in danger. Appropriate training involves recognising and understanding which extinguishers can be used on various types of fires.

Important point

Electrical fires are classified as Class E.

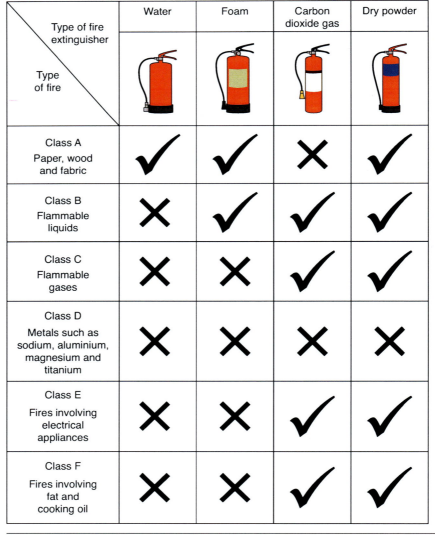

Type of fire extinguisher / Type of fire	Water	Foam	Carbon dioxide gas	Dry powder
Class A Paper, wood and fabric	✔	✔	✘	✔
Class B Flammable liquids	✘	✔	✔	✔
Class C Flammable gases	✘	✘	✔	✔
Class D Metals such as sodium, aluminium, magnesium and titanium	✘	✘	✘	✘
Class E Fires involving electrical appliances	✘	✘	✔	✔
Class F Fires involving fat and cooking oil	✘	✘	✔	✔

Figure 1.33 Electrical fires are classified as Class E

- Water extinguishers should only be used on Class A fires (solid material). They should never be used in any circumstances on electrical or fuel fires.
- CO_2 extinguishers are effective and safe in quenching electrical fires, such as those involving computer equipment. Due to possible asphyxiation, if a CO_2 extinguisher has been operated in a confined space, ensure that the area is ventilated. Never hold the horn when operating a CO_2 extinguisher; although modern versions are fitted with a frost free horn, older versions (made from metal) will cause the operator's hand to be frozen to the horn, causing damage to the skin through what is known as an ice burn.
- Foam extinguishers are particularly good with Class B fires (fuels), because they apply a carpet of foam and deprive the fire of oxygen.
- Dry powder can be used on Class A, B or C fires.

Time out, let's summarise some **key** points!

- Ladders should be used for short durations only
- Any access equipment should be checked before use
- Electrical fires are classified as Class E
- A CO_2 extinguisher is identified through a black band
- Water type extinguishers should only be used on Class A, solid fuel type fires

To review your understanding, try this True or False quiz based on safety systems.

Question 1: Oxygen, fuel and heat form the fire triangle and you need to remove all three to stop a fire.

True	False

Question 2: Anyone can put up scaffolding.

True	False

Question 3: Overhead cables are dangerous because they have a very thin layer of insulation.

True	False

Question 4: All hand-held equipment used on a construction site must be powered through a 110 V supply.

True	False

Question 5: Only serious accidents have to be recorded.

True	False

Five of the Six Pack Regulations

Match the meaning of the following by placing the appropriate number next to the letter in the box provided. The first one has been done for you.

| Management of Health and Safety at Work Regulations 1999 A | A = 2 | A regulation that ensures an employer provides proper work-based equipment (1) |

| Provision and Use of Work Equipment Regulations 1998 B | B = | A regulation that ensures employers provide safe systems of work (2) |

| Manual Handling Operations Regulations 1992 C | C = | A regulation that ensures that all workplaces are safe (3) |

| Workplace (Health, Safety and Welfare) Regulations 1992 D | D = | A regulation that ensures that when hazards cannot be removed PPE is provided (4) |

| Personal Protective Equipment at Work Regulations 1992 E | E = | A regulation that ensures employers train people how to lift loads safely (5) |

Workplace hazards and good working practices

Match the meaning of the following by placing the appropriate number next to the letter in the box provided. The first one has been done for you.

Main cause of accidents in the construction industry A	A = 3	Raise the alarm by warning others and stand guard near the hazard (1)
Main cause of deaths in the construction industry B	B =	Dangerous material that not made any more but is still fitted to some buildings. Dangerous if disturbed (2)
Good husbandry practices C	C =	Slips, trips and falls (3)
Asbestos D	D =	Working at height (4)
Barrier cream and PPE can prevent this skin condition E	E =	Daily clean up and ensuring that tools and equipment are maintained correctly (5)
You notice that some oil has spilt on the floor. What should you do? F	F =	Dermatitis (6)

Workplace procedures

Match the meaning of the following by placing the appropriate number next to the letter in the box provided. The first one has been done for you.

A

A = 5

Even an incident that did not result in harm must be reported in order to learn from the event
(1)

Near miss
B

B =

Type of fire extinguisher that is OK to use with electrical fires
(2)

Recovery position
C

C =

Put an unconscious casualty in this first aid posture
(3)

Assembly point
D

D =

Report here in an emergency
(4)

CO_2 Extinguisher
E

E =

Use a sack trolley when carrying heavy loads
(5)

Health and Safety word search

Find the following hidden words:

HEALTH AND SAFETY, ENABLING, STATUTORY, SIX PACK, HUSBANDRY, PPE, MANUAL HANDLING. MANDATORY SIGN, PROHIBITION, COSHH, ASBESTOS, RCD, REDUCED VOLTAGE, PAT, LADDERS, CLASS E, RIDDOR

H	E	A	L	T	H	A	N	D	S	A	F	E	T	Y	U	H	S
U	R	E	A	N	O	H	S	M	E	M	I	O	N	E	D	C	R
S	I	R	Q	U	S	L	E	S	K	C	A	P	X	I	S	G	E
B	D	C	L	O	I	D	E	R	E	C	G	E	K	A	N	H	D
A	D	G	C	W	O	R	E	A	S	S	E	A	N	R	M	R	D
N	O	N	O	E	I	E	L	E	E	Q	T	R	U	R	I	E	A
D	R	I	S	F	L	W	A	S	B	E	S	T	O	S	N	S	L
R	Y	L	H	T	H	C	I	E	J	S	K	H	S	T	O	I	I
Y	G	B	H	L	O	F	P	I	N	R	I	N	C	I	R	S	L
A	L	A	L	A	E	P	R	O	H	I	B	I	T	I	O	N	K
W	E	N	I	S	T	A	T	U	T	O	R	Y	E	C	I	A	L
S	P	E	S	G	N	I	L	D	N	A	H	L	A	U	N	A	M
A	S	A	R	E	D	U	C	E	D	V	O	L	T	A	G	E	H
A	L	H	E	N	A	B	L	I	N	G	O	F	A	C	T	A	P
C	E	T	A	L	N	G	I	S	Y	R	O	T	A	D	N	A	M

Test your knowledge

1. Know the fundamental aspects of health and safety legislation that apply to electrical installation.

1. The Health and Safety at Work Act 1974:
    a. only affects employers'
    b. only affects employees
    c. only affects customers'
    d. affects everyone on site

2. The Health and Safety at Work Act 1974:
    a. is an act of law
    b. is not an act of law
    c. is non-statutory
    d. does not apply to apprentices

3. Who should make sure that a workplace is safe?
    a. employers
    b. employees
    c. customers
    d. visitors

4. Who should always follow instruction and their training?
    a. passers-by
    b. employees
    c. customers
    d. clients

5. Other regulations can be passed to sit underneath the Health and Safety Act 1974. This is because it an:
    a. enabling act
    b. allowing act
    c. empowering act
    d. assisting act

2. Know how to recognise and respond to hazardous situations.

6. What action should you take if you suspected that a building contains asbestos?
    a. carry on regardless
    b. stop work and wear the correct PPE
    c. stop work and raise the alarm
    d. finish what you are doing then get help

7. Which of the following signs says, 'You must NOT'?
   a. information sign
   b. danger sign
   c. prohibition sign
   d. mandatory sign

8. Which of the following signs says, 'You must do'?
   a. information sign
   b. danger sign
   c. prohibition sign
   d. mandatory sign

9. What kind of sign is this?
   a. safe condition
   b. danger sign
   c. prohibition sign
   d. mandatory sign

10. Which of the following process controls substances?
    a. risk assessment
    b. RIDDOR
    c. COSHH
    d. Permit to work

11. Although nobody was hurt in a workplace incident what kind of report must be raised?
    a. near miss
    b. dangerous occurrence
    c. accident report
    d. first aid

12. When employers carry out a risk assessment, the aim is to seek out any:
    a. lazy workers
    b. hazards
    c. stupid behaviour
    d. shocking behaviour

3. Know the basic safe working procedures.

13. Who should provide PPE?
    a. passers-by
    b. employers
    c. customers
    d. clients

14. When supplied, who should wear PPE?
    a. passers-by
    b. employees
    c. customers
    d. clients

15. In dusty environments the best form of protection would be provided by:
    a. a face mask
    b. a hard hat
    c. a respirator
    d. some overalls

16. PPE can:
    a. stop an accident
    b. stop the effect of injuries
    c. reduce the effect of injuries
    d. increase the effect of an accident

17. The abbreviation SWL stands for:
    a. stable weight load
    b. safe working load
    c. stable working load
    d. safe wide load

18. Which of the following is applicable to correct lifting techniques?
    a. reduced voltage supply
    b. RIDDOR
    c. manual handling
    d. risk assessment

19. **TWO** of the following can cause electrocution. But which ones?
    a. someone touches a cable sheath
    b. someone touches a live supply
    c. someone touches insulation
    d. someone becomes part of a circuit

4. Know how to respond to accidents that occur while working.

20. After a serious injury on site who will conduct any subsequent investigation?
    a. police
    b. council
    c. site manager
    d. Health and Safety Executive

21. You find an electrician who is being electrocuted. You should:
    a. grab them
    b. use a metal object to free them from the supply
    c. fill out an accident report
    d. operate an emergency switch

22. Barrier cream or gloves can prevent:
    a. electric shock
    b. dermatitis
    c. accidents
    d. burns

23. An electrician has burnt themselves, what do you do?
    a. apply hot water to the burn
    b. apply still cold water to the burn
    c. apply cold running water to the burn
    d. apply hot running water to the burn

24. You find another electrician who is being electrocuted. You should:
    a. comfort them
    b. shove them out of the way
    c. use a metal brush to isolate them from the supply
    d. use a wooden brush to isolate them from the supply

25. After an electrician has cut himself or herself, a dressing has been put over the wound, but it continues to bleed. You should:
    a. take the dressing off
    b. turn the dressing over
    c. apply a plaster
    d. apply another dressing

26. The safest option to help someone being electrocuted is to:
    a. drop kick them
    b. earth them
    c. shoulder charge them
    d. go for help

27. An unconscious casualty should be placed:
    a. on their back
    b. on their side
    c. on their face
    d. in the recovery position

28. During a fire alarm drill you should report to:
    a. the canteen
    b. the toilets
    c. your work station
    d. your assembly point

29. Details of all accidents should be recorded:
    a. on accident forms
    b. on site diaries
    c. on work diaries
    d. on accident books

5. Know the basic procedures for electrical safety.

30. Which of the following voltage systems would be the safest to use?
    a. 110 V
    b. 55 V
    c. 230 V
    d. Battery powered

31. Reduced voltage system lessens the effect of an electric shock. But why?
    a. the amount of lethal current is reduced
    b. the amount of lethal power is reduced
    c. the amount of lethal resistance is reduced
    d. the amount of lethal impedance is reduced

32. When dealing with a piece of faulty equipment, which is the correct order in which to proceed?
    a. stop using it, lock it away, call for repair, put a faulty label on it
    b. stop using it, call for repair, lock it away, put a faulty label on it
    c. stop using it, call for repair, put a faulty label on it, lock it away
    d. stop using it, put a faulty label on it, lock it away, call for repair

33. A PAT label on a piece of electrical equipment means:
    a. it has not been tested but is safe to use
    b. it has been tested but user check is still required
    c. it has been tested but user check is not required
    d. it has not been tested and is not safe to use

34. What colour is associated with an 110 V supply?
    a. blue
    b. yellow
    c. red
    d. purple

6. Know the methods of safely using access equipment.

35. Never paint a ladder, because:
    a. certain defects will strengthen the ladder
    b. certain defects and weaknesses can be hidden
    c. it will corrode the ladder
    d. it will weaken the ladder

36. When working above ground for long periods of time, the most appropriate piece of equipment to use would be:
    a. a ladder
    b. a trestle scaffold
    c. a mobile scaffold tower
    d. a step ladder

37. Which **TWO** of the following should only be used for tasks less than 30 minutes in duration?
    a. a ladder
    b. a trestle scaffold
    c. a mobile scaffold tower
    d. a step ladder

38. Who can operate elevated work platforms such as a cherry picker?
    a. trained person
    b. senior electrician
    c. authorised person
    d. manufacturer

39. Which of the following access equipment should be used inside an attic?
    a. scaffolding
    b. crawling board
    c. floor boards
    d. milk crates

8. Know fire safety procedures.

40. Three components are necessary for a fire. They are:
    a. fuel, wood, and cardboard
    b. petrol, oxygen and bottled gas
    c. flames, fuel and heat
    d. fuel, oxygen and heat

41. A fire that has started from a waste paper basket is known as:
    a. a gas fire
    b. a liquid fire
    c. a solid fire
    d. an electrical fire

42. A fire that has started from a waste paper basket is classified as:
    a. Class A
    b. Class B
    c. Class C
    d. Class D

43. An electrical fire is classified as:
    a. Class A
    b. Class B
    c. Class E
    d. Class D

44. When tackling a fuel fire, the most appropriate fire extinguisher to use would be:
    a. water
    b. Class A fire extinguisher
    c. argon
    d. foam

45. You witness a chip pan fire. Which **TWO** of the actions listed is most appropriate?
    a. move the chip pan
    b. throw water over it
    c. switch the gas or electric off
    d. throw a damp cloth over it

Unit: ELEC1/01

Chapter 1 checklist

Learning outcome	Assessment criteria – the learner can:	Page number
1. Know the fundamental aspects of health and safety legislation that apply to electrical installation.	1.1 Outline the aims of general health and safety legislation in protecting the workforce and members of the public. 1.2 Recognise the responsibilities of persons under health and safety legislation.	2 2
2. Know how to recognise and respond to hazardous situations.	2.1 List the types of general workplace hazards that may be encountered while at work. 2.2 Recognise the methods that can be used to prevent accidents or dangerous situations occurring during work activities. 2.3 Recognise warning symbols of hazardous substances. 2.4 Identify the general precautions necessary for working with commonly encountered substances in building engineering service. 2.5 State what action to take should a hazardous situation occur while at work. 2.6 Identify the situations where asbestos may be commonly found in the workplace. 2.7 State the sources of information available in raising awareness of the dangers of asbestos. 2.8 State what actions to take should asbestos materials be identified in the workplace.	7 7 13 13 14 16
3. Know the basic safe working procedures.	3.1 State the purpose and application of protective equipment. 3.2 State the procedures for manually handling heavy and bulky items.	17 20
4. Know how to respond to accidents that occur while working.	4.1 Identify the actions that should be taken when an accident or emergency is discovered. 4.2 Outline the procedures for dealing with minor/major injuries that can occur while working. 4.3 Outline the importance for recording accidents and near misses at work. 4.4 State the procedure for recording accidents and near misses at work.	21 22
5. Know the basic procedures for electrical safety.	5.1 Identify the common electrical dangers encountered on construction sites and in dwellings. 5.2 Identify the methods of safely using electrical tools and equipment on site.	27
6. Know the methods of safely using access equipment.	6.1 Outline the situations where it may be necessary to work at height. 6.2 State the safety requirements for the types of equipment used to permit work at heights in the building services industry. 6.3 List the safety checks required to be carried out on access equipment before it is used.	31
8. Know fire safety procedures.	8.1 Outline how combustion occurs. 8.2 State the method for tackling small localised fires that can occur in the workplace.	34 34

Electrical science, principles and technology

EAL Unit ELEC1/08

Learning outcomes

The learner will:

1. Know the fundamental units of measurement used in electrotechnical work.
2. Know the basic principles of electrical circuits.
3. Know basic aspects of diagrams and circuits.

EAL Electrical Installation Work – Level 1. 978-1-138-23206-8

Learning outcome 1.1

Identify basic (SI) units of measurement for general quantities.

Learning outcome 1.2

Outline the SI or derived SI unit for electrical quantities.

How long do we have to finish this job?

How much cable do we have left?

Is that bundle of conduit heavy?

All these types of questions will typically be used on a construction site, but in order to be able to use this kind of information effectively, then everyone involved in the conversation must understand what they represent.

It is very difficult, if not impossible, for any two people to communicate with each other if they do not speak a common language. This is no different in electrical installation, because we need to agree and understand certain elements of nature within our natural surroundings before we can think of carrying out any activity such as installing and testing an electrical system.

The term SI unit is an international system which contains certain base units and also what are known as derived units which are units that have combined one or more of the base units.

Base units include:

- Metre to measure length
- Kilogram for the mass of an object
- Second to measure time
- Ampere for the flow of electric current
- Kelvin to measure temperature
- Candela for luminous intensity (light)
- Mole for the amount of substance in a body

Description

SI units represent an agreed international format used to standardise certain measurements.

Description

The SI unit of length is the metre.

Description

Electrical current is measured in amperes sometimes shortened to amps.

Basic SI units

Quantity	Measure of	Basic unit	Symbol	Comments
Temperature	Hotness or coldness	Kelvin	K	K 0°C = 273 K Absolute zero (0 K) = −273.15 °C
Length	Distance	Metre	m	The metric system has been extended to incorporate many more sub-multiples
Mass	Mass of a particular international prototype made of platinum-iridium and kept at the International Bureau of Weights and Measures	Kilogram	Kg	
Current (I)	Electric current	Ampere	A	The rate at which electricity flows

Quantity	Measure of	Basic unit	Symbol	Comments
Time	Time	Second	S	60 s = 1 min, 60 x 60 s = 1 hour
Luminous intensity	Intensity of light	Candela	I	
Mole	The number of atoms or molecules	Mole	*n*	

Some of these will be familiar to you but others will not. You need to remember that these are all base units, in other words a starting point for a particular aspect. Let us examine these units in greater detail.

Metre

The metre was first used to define measurements but of course is also part of the metric system, which is the system of measurement used in science and engineering in most countries but not America. The metric system also uses sub-multiples such as mm and cm to measure smaller distances as well km, which is used to gauge and represent a much larger quantity.

Mole and mass

The SI unit mole defines how many atoms are contained within an object. To explore this further let us consider the following question.

Which is heavier to lift – a football or a bowling ball?

The answer to the question is that the bowling ball is far heavier, something that anyone who has ever played ten pin bowling can confirm.

But why?

The bowling ball is made from a substance or material that is much more compact with matter or in other words, its material contains a lot more atoms. The mole is therefore the unit which scientists use to measure how many atoms are contained within a material. The term mass, however, refers to the overall amount of matter involved and is measured in kg.

Weight on the other hand, is actually the force of gravity acting on the mass of an object. You can see this relationship through its formula: weight = mass x gravity. Force is a derived unit and its SI unit is the newton, symbol N, named after the scientist Sir Isaac Newton. When we weigh ourselves, we are simply weighing the mass of our bodies having removed gravity from the process because gravity is a constant value and affects all objects and bodies in equal measure. One last point – the mass of an object cannot be altered; the actual amount of substance or atoms is a known quantity. Weight, however, does change; for instance, an object will weigh less on the moon because the effects of gravity is far less, roughly 1/6 the strength of gravity on the earth.

Time

Time is represented by seconds as its base unit, but it is important to remember that time can also be measured in minutes, hours and so on. This is no problem

Description

The SI unit of time is the second.

as long as we remember the difference between each unit of time and we convert when required. For instance:

2 minutes is 120 seconds

1 hour is 60 (minutes) x 60 (seconds) = 3600 seconds.

Ampere

The term ampere represents the amount of electricity that flows in an electrical circuit. Named after André-Marie Ampère (1775–1836) a French mathematician, it is often shortened to amp. A useful way of remembering what this unit represents is to associate the term ampere to an electrical fuse. In other words, to remember the SI unit for electrical current = think fuse (as in a 3 Amp fuse).

Temperature

Description

The SI unit of temperature is the kelvin.

Although 0 degrees is recognised as the freezing point of water, other materials including certain gases freeze or stop moving at temperatures much lower than this. This is where the unit Kelvin was introduced, named after Lord William Kelvin, who declared that the starting point for all types of matter should be absolute zero, which is registered at −273°C. To relate to the melting and freezing point of everyday items however, Fahrenheit – named after the German physicist Daniel Gabriel Fahrenheit – is sometimes used. Much more common however, is Centigrade, a system named after the Swedish astronomer Anders Celsius. Celsius actually means a difference between two temperature ranges such as 0°C to represent the freezing point of water and 100°C to indicate the boiling point of water.

Candela

The candela is simply a measure of light intensity and it can be thought of as candlepower in the same way as early measurement of mechanical power was calculated by horsepower.

Luminous intensity

Luminous intensity is the amount of visible light and power given off from a light source.

Derived units

Description

A derived unit uses one or more of the SI base units.

All the units examined so far can be combined to form what are known as derived units and a good example is the measurement of acceleration, which is defined as how fast an object changes speed. For example, if a car is travelling at 5 miles per hour (mph) but then accelerates to 15 miles an hour then the actual acceleration equates to the difference between the two: 15 − 5 = 10 mph. Acceleration is actually measured in metres per second^2 (m/s^2) and therefore combines two base units: metres regarding distance and seconds for time. Because Fahrenheit and Centigrade are based on the Kelvin system, they too are both derived units.

Derived units are also associated with certain mathematical measurements such as:

* Area
* Volume
* Density

Table 2.1 Basic SI units

Quantity	Measure of	Basic unit	Symbol	Notes
Area	Length × length	Metre squared	m^2	
Current (I)	Electric current	Ampere	A	
Energy	Ability to do work	Joule	J	Joule is a very small unit 3.6×10^6 J × 1 kWh
Force	The effect on a body	Newton	N	
Frequency	Number of cycles	Hertz	Hz	Mains frequency is 50 Hz
Length	Distance	Metre	m	
Mass	Amount of material	Kilogram	Kg	One metric tonne = 1000 kg
Magnetic flux Φ	Magnetic energy	Weber	Wb	
Magnetic flux density B	Number of lines of magnetic flux	Tesla	T	
Potential or pressure	Voltage	Volt	V	
Period T	Time taken to complete one cycle	Second	s	The 50 Hz mains supply has a period of 20 ms
Power	Rate of doing work	Watt	W	
Resistance	Opposition to current flow	Ohm	Ω	
Resistivity	Resistance of a sample piece of material	Ohm metre	ρ	Resistivity of copper is $17.5 = 10^{-9}$ Ωm
Temperature	Hotness or coldness	Kelvin	K	0°C = 273 K. A change of 1 K is the same as 1°C
Time	Time	Second	s	60 s = 1 min 60 min = 1 h
Weight	Force exerted by a mass	Kilogram	kg	1000 kg = 1 tonne
Electric charge	Charge transported by a constant current of one ampere in one second	Coulomb	C	Charge of approximately 6.241×1018 electrons

Note: A more detailed description may be found in this chapter.

For example, to calculate the area of an object, we simply multiply its length by its width. Area is indicated by square metres or m^2, because we are effectively multiplying two sides.

Example: if the length of a rug is 5 m and its width is 4 m, calculate its area?

Area = length x width

Area = 5 x 4

Area = 20 m^2

To calculate the area of a right angle triangle we use the same method but it does differ slightly. This type of triangle can be thought of as half a square or rectangle, therefore our formula becomes:

$$A = \frac{b \times h}{2}$$

If the height of a triangle is 15 cm and its base is 8 cm long, calculate its area:

Top tip

Always write out an equation in steps.

Step 1: formula

Step 2: put in your numbers

Step 3: calculate the answer including its unit

$$\text{Area} = \frac{b \times h}{2}$$

$$\text{Area} = \frac{8 \times 15}{2}$$

$$\text{Area} = 60 \text{ cm}^2$$

Description

Area is measured in square metres m².

Figure 2.1 Cables tend to be measured in mm², which is a measure of its cross sectional area

Description

Volume measures three dimensions (m³) and can be thought of as the amount of cubic space contained inside an object.

Theory into practice

Later on in this chapter, we will discuss how ring circuits are used to supply electrical power to socket outlets but ring circuits are **not** permitted to supply any room bigger than (>) 100 m² in area. This is a perfect example of why we need to understand basic mathematics so that we can apply it to electrical installation practice.

Apart from the example above, electrical installation employs another measurement involving area, which is used to establish what size cable is required to carry a particular electrical current. The science behind this is very important for an electrician because undersized conductors cannot cope with the heat that is generated, which in turn will cause its surrounding insulation to melt. Cable size actually refers to the cross sectional area of the conductors with most cables being measured in mm² (Figure 2.1).

Volume is a measure of how much space is occupied within an object but with regard to three dimensions. When assessing this space we must consider an object's length, width and height, which is why volume is represented by metres cubed, m³.

We can also use volume to consider how much space is contained in an object such as working out that the container shown in Figure 2.2 can hold 8 cups.

Figure 2.2 Volume is measured in m³

Worked example

Calculate the volume of the cube shown in Figure 2.3 given that its dimensions are:

Length: 20 cm

Width: 20 cm

Height: 20 cm

Volume = length x width x height

Volume = 20 x 20 x 20

Volume = 8000 cm³

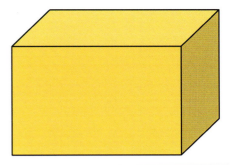

Figure 2.3 To work out the volume of a cube, simply multiply its length, width and its height

Fluids

When we store fluids, we tend not to use the term volume but instead use the term capacity.

Although the basic formula for volume is always the same: area (length x width) multiplied by its height. When fluids are stored in cylinders (Figure 2.4), we need to account for the fact that their base is cylindrical (round) therefore we use the formula: $a = \pi r^2$.

A lot of mathematical formula include π (Pi), but what is π?

If you measure the circumference and diameter of any round object and then divide the value of the circumference by its diameter then you obtain π, whose value to 3 significant figures is 3.142. This value is constant and will not change even if you compare the measured circumference and diameter of the earth (Figure 2.5) or a two pence coin (Figure 2.6).

Figure 2.4 To calculate the area of a circle or the base of a cylinder then the formula involved is $a = \pi r^2$

Figure 2.5 Dividing the earth's circumference by its diameter will produce the value π (3.142)

Figure 2.6 Dividing the circumference of a two pence piece by its diameter will produce the value π (3.142)

Worked example

Calculate the volume of the cube shown in Figure 2.3 given that its dimensions are:

Length: 20 cm

Calculate the volume of the cylinder if its height is 5 cm and the radius of its base is 3 cm.

Volume = (area of a circle) × (height)

Volume = (p × *radius²*) × (height)

Volume = (3.142 × 3²) × (5)

Volume = 28.278 × 5

Volume = 141.39 cm³

Fluid is actually measured in millilitres or litres and its worth remembering that 1 m³ is equal to 1000 litres or 1 k litres.

We have already stated that the unit of electric current is the ampere but an electrician will need to understand and use other electrical quantities such as:

- Voltage. Measured in volts, this is the difference in potential between two points. For instance, the difference between the line and neutral conductor in a domestic single-phase supply is 230 V. It might also be useful for you to think of voltage as electrical pressure.
- Electric charge. Certain items can hold or store electricity and the amount of charge is measured in coulombs. The coulomb is equivalent to one ampere per second.
- Resistance. Resistance is a measure of how much opposition there is to an electrical current. If an electrical conductor is very long, then its resistance will increase accordingly. The SI unit of electrical resistance is the ohm (Ω).
- Power. The amount of electricity consumed over a particular time is a measure of electrical power and its SI unit is the watt.

In other words, power can be defined as the rate of doing work (energy); bearing in mind that in an electrical circuit this energy is actually heat being produced. A useful way of remembering what unit is associated with power is to associate this term with an electrical light bulb because they too are rated in watts.

Description

Potential difference is measured in volts (V).

Description

Resistance is measured in ohms (Ω).

Description

Electrical power is measured in watts (W).

SI Units activity A

This activity contains two sets of matching terms and each matching pair should be given the same number. The first one has been done for you: 'Ampere = 1' matches 'Electrical current = 1'

Ampere = 1	Voltage = 2		Electrical current = 1	Potential difference =
Power =	Energy =		Joule = 3	Watt = 4
Length = 5	Time = 6		Resistance =	Degrees Kelvin =
Ohms = 7	Temperature = 8		Metre =	Second =
Coulomb =	Mass =		Electrical charge = 9	Kg = 10

SI Units activity B

Put an X in the correct box that identifies if the terms are an SI base or derived unit. One has been done for you.

Put an X in the correct box	SI base unit	SI derived unit	Electrical unit
Time			
Electric current (amperes)	x		x
Coulomb			
Length			
Area			
Voltage			
Power			
Mole			
Mass			
Temperature			
Resistance			

Learning outcome 1.3

List the common multiples and sub-multiples used within electrotechnical work.

We have already discussed that measurements of electrical quantities include both base units such as ampere (flow of electricity) and derived units such as coulomb (electrical charge). Engineers tend to represent numbers in either big multiples or small sub-multiples of 10 because the numbers involved can be incredibly big or incredibly small in value and therefore it is easier sometimes to represent them through another convention (system).

A system known as scientific notation is used because it allows you to handle very large or very small numbers with ease. For example, instead of writing 0.0000000039, we would write 3.9×10^{-9}. The minus sign indicating that the number has to be divided by the subsequent number of zeros.

In electrical installation and other forms of engineering, we tend to use another system, which is very similar in design, called engineering notation. Engineering notation only uses powers of ten that are multiples of 3. For example, to represent very small values, 10^{-3}, 10^{-6}, 10^{-9}, 10^{-12} are used. Going the other way, very large numbers are represented by 10^3, 10^6, 10^9, 10^{12}.

Let us examine this further

If 1 amp is divided into a thousand bits this effectively is a fraction:

1/1000 A

This fraction can also be written as 0.001 A, when expressed as a decimal. All that has happened is that the decimal point has been moved 3 times to the left of the figure 1.

Description

1/1000 is represented by the sub-multiple milli.

Equally, this can also be written as 1 ×10⁻³ A, the minus sign indicating that the number has to be divided by a factor of 3.

This means that all the figures below represent the same value:

$1/1000 = 0.001 = 1 \times 10^{-3}$

Table 2.2 Symbols, multiples and sub-multiples for use with SI units

Prefix	Symbol	Multiplication factor		
Mega	M	$\times 10^{6}$	or	$\times 1,000,000$
Kilo	k	$\times 10^{3}$	or	$\times 1000$
Hecto	h	$\times 10^{2}$	or	$\times 100$
Deca	da	$\times 10$	or	$\times 10$
Deci	d	$\times 10^{-1}$	or	$\div 10$
Centi	c	$\times 10^{-2}$	or	$\div 100$
Milli	m	$\times 10^{-3}$	or	$\div 1000$
Micro	μ	$\times 10^{-6}$	or	$\div 1,000,000$
Nano	n	$\times 10^{-9}$	or	$\div 1,000,000,000$
Pico	ρ	$\times 10^{-12}$	or	$\div 1,000,000,000,000$

Description

1,000,000 is represented by the multiple mega.

Larger numbers such as 1000 can be written as 1×10^{3} (how many zeroes in a thousand = 3).

These sort of techniques are very handy for an electrician because electrical test equipment uses values which are registered in millions. For instance, an insulation resistance meter tests how effective a cable's insulation is and the healthier the insulation, the higher the value. Values for a brand new circuit could be in excess of 100,000,000 Ω but is easier when it is represented as 1×10^{9} (count the number of zeros involved = 9).

For the sake of convenience, engineering notation can also be represented with various prefixes, as indicated in Table 2.2.

This means that yet again all the figures below actually represent the same value:

$1/1000 \text{ A} = 0.001 \text{ A} = 1 \times 10^{-3} \text{ A} = 1 \text{ mA}$

If I decided to divide the number one (1) by a million then I could write it as:

Description

1/1000000 is represented by the sub-multiple micro.

- 1/1000000 A (1 divided by a million) or
- 1×10^{-6} A (1 divided by a million) or
- 0.000001 A (decimal point has been moved 6 times to the left)

Alternatively, it is far easier to use the prefix μ (micro).

Look at the example below and decide for yourself, which is easier to write?

0.000001 A, 1/1000000 A or 1 μA.

When using these values in calculations it is often useful to realise how prefixes relate to each other. For instance, a number can appear bigger because you are representing a smaller unit through its prefix.

Description

1000 is represented by the multiple kilo.

For instance:

There are 1000 mm in every metre: 1 m = 1000 mm

There are also 100 cm in every metre: 1 m = 100 cm

1000 and 100 are obviously bigger than 1 and this is because they are representing mm and cm respectively; in other words, far smaller units when comparing them to a metre.

In electrical terms, therefore:

0.03 A is also equal to 30 mA

But why?

This is because we have converted from a big unit (amps) to a smaller unit (milli amps). The number becomes bigger by a factor of 3 (milli = 10^{-3}).

Let us go the other way and convert from a small unit to a bigger unit.

Convert from 3 µA to amps (A)

The answer is 0.000003 A, because we have converted from a small unit (3 µA) to a bigger unit – amps. The decimal point has been moved six times to make the number smaller by a factor of 6.

Try to work out if the following example is true:

30 µA is equal to 0.03 mA which is equal to 0.00003 A – Ask your tutor!

Multiples, sub-multiples and prefixes activity

This activity contains two sets of matching terms and each matching pair should be given the same number. One has been done for you: 'Kilo = 5' matches '1 x 10^3 1000 = 5'

milli = 1	micro = 2		1 x 10^3 1/1000 =	1 x 10^6 1/1000000 =
nano =	pico =		1 x 10^{-9} 1/1000000000 = 3	1 x 10^{-12} 1/1000000000000 = 4
Kilo = 5	Mega = 6		1 x 10^3 1000 =	1 x 10^6 1000000 =
Terra =	Giga =		1 x 10^{12} 1000000000000 = 8	1 x 10^9 10000000000 = 7

Time out, let's summarise some **key** points!

- Certain units of measure are used to meet an international agreement (SI) on how units are represented

- For instance, when measuring out the length of a room then in the majority of cases, it would be shown in metres. However, other sub-multiples such as cm and mm are also used, for example when converting actual distances in metres on to a scale diagram

- Other formulas an electrician will use include: area (m^2) and volume, which measures the amount of space, that is occupied within a 3 dimensional shape (m^3)

- Scientific notation is a system that can be used to represent very large or small numbers often through prefixes such as kilo, which is used to represent 1000

To review your understanding try this True or False quiz based on SI units.

Question 1: The SI unit for energy is the Watt.

True	False

Question 2: Milli is a submultiple and can be thought of as 1 ÷1000 (1/1000).

True	False

Question 3: Kilo is a prefix, which represents 1000, or 1×10^3.

True	False

Learning outcome 2.1

Determine basic electrical quantities using Ohm's law.

Of all the laws that an electrician needs to learn, one of the most important is Ohm's law.

The actual law was defined by a German scientist named George Ohm, who described the relationship between:

- Voltage
- Current
- Resistance

These units have already been described earlier but another important aspect is that we understand how they relate to one another.

Sometimes a comparison can be made between two different things in order to explain how something works and in the case of electricity, this comparison is known through the Electricity-Water Analogy (Figure 2.7)

Tackle a difficult word: analogy

Explaining something by comparing it to something else

Figure 2.7 Water is said to flow through a pipe. So too electricity, which flows through a conductor

The flow of electricity (Figure 2.8) is equivalent to the flow of water (Figure 2.9).

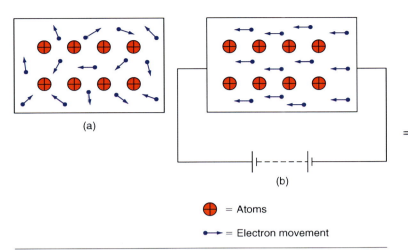

(a)

(b)

● = Atoms

•—→ = Electron movement

Figure 2.8 Electrons are negatively charged and move towards a positive electric force

Figure 2.9 Electrical current is measured in amperes, sometimes shortened to amps and flows through a conductor just like water

Voltage (Figure 2.10) is equal to water pressure (Figure 2.11) and this is why it is sometimes called electrical pressure.

Various engineering systems use pumps to produce pressure so that oil, petrol or water is pumped around the system. An electric battery is very similar and produces electrical pressure by creating a difference in potential between the positive and negative plate. Measuring 9 V across a battery is actually a 9 V difference in potential between the positive and negative terminals.

Important point

Voltage is a measure of potential difference but can also be thought of as electrical pressure.

Figure 2.10 Voltage can be thought of as electrical pressure

Figure 2.11 Air pressure is used to drive pneumatic tools. Electric pressure is used to push electrical current which is then used to operate electrical equipment

The electrical cable (Figure 2.12) can be compared to a water pipe (Figure 2.13).

A large pipe can carry more water than a smaller one. This is true of an electrical cable, because a big cable will have less resistance. Less resistance will mean more current can flow.

Let us examine this through a further analogy.

Important point

A large cable can carry more electricity than a smaller one. This is because it has less resistance.

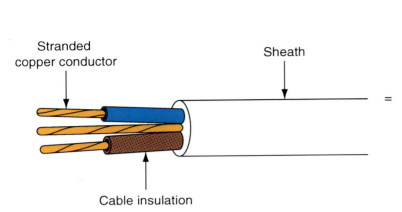

Figure 2.12 A large cable can carry more electrical current than a small one

Figure 2.13 A large hose pipe can carry more water than a small one

Let us pretend that you are an electrical current and you have heard that there are two shops that are giving away Playstations and televisions. Both shops have door staff, but which one would you chose to enter? Shop A? (Figure 2.14a) or Shop B? (Figure 2.14b)

Shop A

Shop B

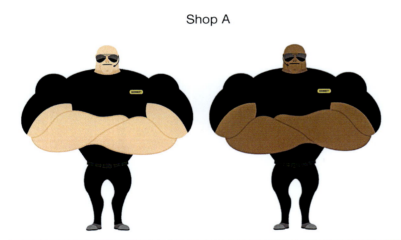

Figure 2.14 (a) More door staff is like having more resistance

Figure 2.14 (b) Less door staff is like having less resistance

You have more chance of entering Shop B because there is only one door staff on duty. Think of this in terms that there is less resistance to you entering. An electrical current will do the same and follow the path of least resistance for itself.

Shop A is equal to a small cable: a lot of resistance to the flow of current (more door staff)

Shop B is equal to a larger cable: less resistance to the flow of current (less door staff)

If Ohm's law were an Olympic event such as hurdles, then:

- The runner would represent the current flow of a circuit (amperes)
- The track would represent the cable
- The hurdles would represent the resistance
- The actual pressure on the runner to win could even represent the voltage (electrical pressure)

Figure 2.15 Hurdles provide physical resistance to the runner. Resistance provides physical resistance to electrical current

Ohm's law triangle

Some very important facts regarding the relationship between voltage, electrical current and resistance can be seen from the diagram below. Consider the following:

If a battery is applied across a circuit as seen in Figure 2.16, as long as the circuit is complete then the circuit will become live. This means that:

• The current flows through the conductors and resistors
• The battery voltage appears across the resistors

The actual resistance of the circuit opposes electrical current, in other words it acts against it.

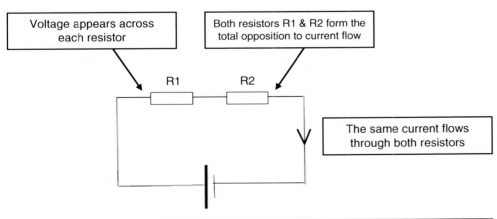

Figure 2.16 Electrical current flows through a component, voltage appears across it

Important formula coming up

Ohm's law can also be represented by a triangle as shown in Figure 2.17.

Put your finger on the element you want to find in order to extract all three formulae.

Ohm's law can be defined through a triangle:

$$\frac{V}{I \times R}$$

Figure 2.17

$$R = \frac{V}{I}$$

$$V = I \times R$$

$$I = \frac{V}{R}$$

Top tip

Always write down calculations in three stages, as shown in the worked example.

Worked example

1) The heating element inside an electric kettle needs a voltage of 230 volts and 10 amps of current in order to operate. Work out the resistance of the heating element.

$$R = \frac{V}{I}$$

$$R = \frac{230}{10}$$

$$R = 23 \ \Omega$$

2) An electric fire needs 12 amps of current to work. If its heating element has a resistance of 20 Ω, work out the value of the supply voltage.

$$V = I \times R$$

$$V = 12 \times 20$$

$$V = 240 \text{ V}$$

Why not attempt the next question yourself?

3) An electric toaster is supplied with a domestic voltage of 230 volts and needs 9 amps to produce enough heat to toast some bread. Work out the resistance of its heating element.

Time out, let's summarise some **key** points!

- Understanding Ohm's law is very important for an electrician. All the elements can be compared to how water flows through a pipe
- The flow of electrical current is measured in amperes; this is sometimes shortened to amps
- It is vital to establish that water flows through a pipe; electrical current is no different and flows through a conductor

- Voltage or potential difference can be thought of as electrical pressure that forces the current around the circuit.
- Electrical resistance (measured in Ω) can be compared with the size of a water pipe. A big pipe can carry more water than a smaller one.
- This is no different to an electrical cable because, a big cable has less resistance and therefore can carry more electrical current.
- In other words, resistance opposes electrical current.

To review your understanding try this True or False quiz based on Ohm's law.

Question 1: When a circuit becomes live, current flows through a component and voltage appears across it.

True	False

Question 2: Resistance helps electric current.

True	False

Question 3: This is the Ohm's law triangle.

True	False

Figure 2.18

Learning outcome 2.5

Outline the basic principles of direct and alternating current.

Electricity comes in two flavours and they are known as direct current and alternating current. Electrical equipment can be designed to operate on both but only one type is used in the transmission and distribution of electricity across many thousands of miles. Why?

To answer this question we must go on an electrical journey!

A typical power station uses fossil fuels such as oil, coal and gas as well as nuclear energy to heat up a large amount of water to boiling point in order to create steam. The steam is then directed towards large turbines, which drive generators that produce an alternating electrical supply.

A direct current (d.c.) supply is basically electricity that flows in one direction and its value remains constant. Batteries produce d.c. to operate such items as mobile phones and laptops. Car batteries also require a d.c. current in order to remain charged so that they can be used to restart the engine once the car has been parked and the engine switched off.

Important point

There are two types of electrical current – direct current (d.c.) and alternating current (a.c.).

All generators produce alternating current (a.c.) but other devices produce direct current. For example, most cars are fitted with an alternator and, whilst the engine is turning over, the alternator will produce alternating current. The alternator electrical supply, however, will be fed to some diodes, which will convert or rectify the alternating current to direct current. Even very early pedal bikes were fitted with a device called a dynamo, this time using a mechanical device called a commutator to produce a direct current electrical supply. The dynamo generated enough electricity whilst the bike was being pedalled to power up certain lights.

Electricity is also needed in domestic properties (houses), commercial (offices, shops, restaurants) and industrial (factories) locations. Therefore, the sheer range of locations that require electricity is vast and, given that electricity has to be transmitted and distributed over thousands of miles, the amount of current needed is huge.

If the electricity supplied were direct current in nature, then extremely large conductors would be required in order to carry this enormous current. Direct current would therefore be a very expensive way of distributing electricity.

However, a way was found of transmitting and distributing electricity but in the process actually reducing the size of conductors involved. This system relies very heavily on a particular device known as a transformer but, for reasons that will be discussed later, this device only operates with alternating current. An alternating current network was therefore developed because it is a far more efficient and economical system and it is distributed by overhead (Figure 2.19) or underground power lines.

Figure 2.19 Transformers are used in the transmission and distribution of electricity

Although Ohm's law is very important for an electrician, it does not apply with transformers.

Normally in an electrical circuit if the resistance stays the same, any increase in voltage will see a matching increase in current.

A transformer contains two coils, which are not linked electrically but are joined by an alternating magnetic field. When an electric current is passed through one coil, it induces current in the other.

Referring to Figure 2.20, the input coil is known as the primary coil and the output is known as the secondary. It might be helpful to use an analogy that the magnetic field that links them is similar to water rippling between two stones,

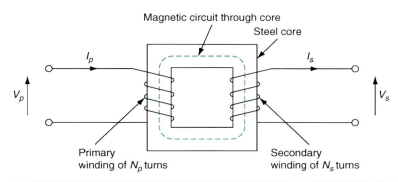

Figure 2.20 A transformer's input coil is known as Primary. It's output coil is referred to as its Secondary

Figure 2.21 A transformer operates through magnetism, which ripples between the primary and secondary coils

which is represented in Figure 2.21. The correct term for its operating principle is mutual induction.

Important formula coming up

Another important formula is known as the main formula for power for an electrician:

Power = Voltage x Current

The amount of power contained inside this rippling magnetic field cannot be changed. This means that if we increase the amount of voltage, to keep the amount of power the same, then the circuit current will actually reduce. In fact, doubling the voltage will halve the current.

Power (cannot be changed) = Voltage x Current

This is why transformers only work with alternating current, the actual ripple we spoke of earlier using the two stones as an example (Fig 2.21) needs to be changing all the time to keep the magnetic field going.

Transformers are therefore used to boost voltages up to huge amounts such as 400 kV (400,000 volts). Not only will reduced current levels mean smaller cables, but the amount of heat given off is also far less which makes this system far more efficient.

Important point

A domestic mains supply provides alternating current (a.c.).

The electricity associated with a mains supply and socket outlets (Figure 2.22) is therefore alternating current, but other items such as batteries operate with direct current at a much-reduced value.

Figure 2.22 Mains electricity provides alternating current

Important point

Laptop and mobile phones require direct current (d.c.).

This is why your mobile phone and laptop adapters contain two key devices inside. The first is a small transformer so that the voltage can be reduced down to 12 V. The other is a diode, an item that only allows current to flow one way – in other words changes alternating current to direct current and is known as a rectifier.

Learning outcome 2.2

List the effects of an electric current.

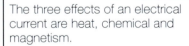

Description

The three effects of an electrical current are heat, chemical and magnetism.

An electric current can produce three effects within a circuit and they are:

- Heating
- Magnetic
- Chemical

Heating effect
When an electric current flows this is basically a flow of energy and this energy is turned into heat. Different domestic appliances use this heat to operate and we will now examine how.

The iron
A modern iron (Figure 2.23) produces heat but also heats up water to produce steam in order to make the ironing process easier. The heat is controlled by a thermostat, which varies the element's resistance. Reducing the thermostat's resistance for instance, increases the current available and the amount of heat generated. This type of control and flexibility is needed to remove wrinkles from a wide range of materials.

Figure 2.23 A steam iron operates through heat (thermal effect)

The fuse

A fuse is classified as a type of circuit protective device and Figure 2.24 shows a cartridge type with a glass tube. The rating of a fuse is designed to be at least equal to the value of the normal circuit current. During a fault, however, the value of the current becomes excessive and the amount of heat produced is too much for the fuse wire inside which melts or blows to isolate the circuit. A fuse therefore operates through a thermal effect (heat) and this is how it is purposely designed to be the weakest part of an electric circuit.

Description

A fuse operates through a thermal effect (heat).

Fuse element

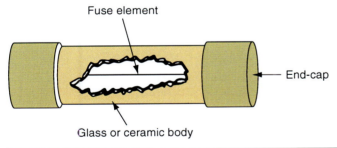

End-cap

Glass or ceramic body

Figure 2.24 A fuse is designed to be the weakest part of a circuit. Excess current produces excess heat and melts the fuse wire

Chemical

A battery such as that shown in Figure 2.25, includes several cells all connected together and uses two different types of metals to act as electrodes which are commonly known as + (positive) and - (negative). These electrodes are covered with a solution called an electrolyte and when an electric current is passed through it, a chemical reaction causes negative particles to be collected on one electrode and positive on the other, in other words, producing a potential difference.

Description

A battery produces an e.m.f through a chemical effect.

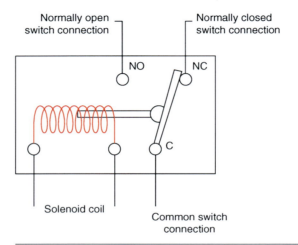

Figure 2.25 A battery operates through an electrochemical effect

Magnetism

The third effect produced when an electric current flows through a conductor is magnetism. If you remember we have previously discussed magnetic fields, as Figure 2.19 and Figure 2.20 were used to describe how a transformer uses magnetism to mutually link two coils.

Coils are also known as inductors and by definition are regarded as electromagnets because they generate a magnetic field whenever electricity is passed through them. Electromagnets have many uses and we will consider each in turn.

Relay

A relay comprises of two main components: an electromagnet called a solenoid (Figure 2.26) and a set of contacts. When an electric current flows through the solenoid a magnetic field is produced to shut close the contacts. Relays have many applications such as:

- Electric bells
- Vehicle central locking
- Alarm systems

Other devices that use magnetism to operate include circuit breakers, designed to operate by sensing when a fault current is produced. This fault current will be extremely large and will in turn create an equally large magnetic field. It is this magnetic field that is used to trip the device very quickly in order to isolate the circuit.

In a healthy electrical circuit, electricity is delivered to a circuit through the line conductor and is returned to the supply by the neutral conductor. This is why electricity can be thought of as ET: to operate correctly it must find its way home!

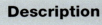

Description

An electromagnetic device only produces magnetism when an electric current flows through it.

Normally open switch connection

Normally closed switch connection

NO

NC

C

Solenoid coil

Common switch connection

Figure 2.26 A relay includes an electromagnetic device known as a solenoid and then an arrangement of contacts

Any fuse or circuit breaker will blow or trip because the fault current concerned will be many times the size of the normal circuit current. If this fault current if passed down the earthing system, another device will operate far quicker. This device is called a residual current device, sometimes abbreviated to RCD, and it, too, operates through magnetism.

Power triangle

Electrical power is defined as the amount of energy generated or used over a given time. In an electric circuit, this energy becomes heat or light, and is dependent on how much current is flowing (amps), which in turn depends on electrical pressure (voltage).

We examined the power formula earlier but it can also be shown through the Power triangle as seen below. As stated previously, this formula is very important for an electrician since it is used in the very first steps when designing any electrical system in order to work out how much electrical current an installation requires.

First, let us remind ourselves of the units involved.

Figure 2.27

Power = watts

Symbol: P

Voltage = volts

Symbol: V

Current = amperes

Symbol: I

You can extract three possible formulas from the triangle, simply by putting your finger on to each letter in turn.

The first is shown below, covering the letter I to extract the formula for finding electrical current:

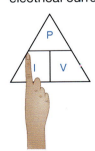

Figure 2.28

$$I = \frac{P}{V}$$

Why not have a go at finding the formula for P (power): P =

Then apply the same finger to find the formula for Voltage: V =

Worked example

Electrical power calculations

1) If a resistor has 10 volts developed across it and 10 amps flowing through it, calculate the circuit power?

P = V x I

P = 10 x 10

P = 100 W

2) If a circuit is supplied with 230 V and produces 3,000 watts of power, calculate the current flowing in the circuit.

$$I = \frac{P}{V}$$

$$I = \frac{3000}{230}$$

I = 13.04 amps

Why not attempt the next question yourself?

3) If a cable drops 4 V along its length and has 25 A of current flowing through it, how much power is produced?

Time out, let's summarise some **key** points!

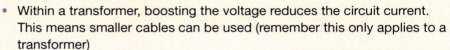

- There are two types of electricity, both are used to power equipment and both are potentially dangerous
- Laptop or mobile phones require direct current (d.c.) but a domestic mains supply is associated with alternating current (a.c.)
- Transformers are used to both boost voltages up and step voltages down but can only do so when supplied with alternating current
- Within a transformer, boosting the voltage reduces the circuit current. This means smaller cables can be used (remember this only applies to a transformer)
- This relationship is shown through the main formula for power for an electrician: P = V x I
- When an electric current flows through a circuit, it produces three effects: heat, magnetism and a chemical reaction

To review your understanding try this True or False quiz based on electrical supplies.

Question 1: A domestic mains supply is associated with direct current (d.c.).

True	False

Question 2: A transformer only works when supplied with an a.c. supply.

True	False

Question 3: We can extract the main formula for power for an electrician by using this triangle.

True	False

Electric circuits

It was established earlier that the amount of electricity that flows is represented by the term ampere. However, it is very important to understand that an electric current is produced by a certain particle inside an atom. Representing the smallest part that actually makes matter, the atom is broken down into various elements as shown in Figure 2.29. The electron is what actually creates electricity and it has a negative charge. The other two elements are called the proton (positive charge) and a neutron (no charge). If an atom has an equal number of electrons, protons and neutrons then the atom is said to be balanced. This is because their charges all cancel out.

Description

Electricity is caused by the movement of negatively charged electrons.

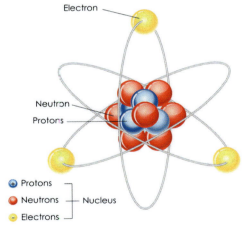

Figure 2.29 Electrical current is produced by the movement of tiny particles called electrons, which are associated with a negative charge

When a positive energy source such as a voltage or potential difference is applied across a conducting material, it gives certain electrons known as valance electrons some energy, which allows them to move or hop to a different position or level inside the atom. The movement of valance electrons is what causes electrical current.

Voltage therefore actually attracts current around a circuit rather than pushing it around. Some materials contain a lot of these electrons and are kept in place by a weak bind. This means they are easily influenced and because of how they are constructed, they conduct electricity easily. They are classified as conductors and include materials such as metals and carbon.

Other materials contain only a few electrons but these electrons are kept in place by a very strong bind. Because they do not easily conduct electricity, they are classified as insulators. Insulators can therefore be used to protect people from the effects of electric shock. Typical insulating materials include plastic, wood, paper and even air.

Both conductors and insulators are used extensively in electrical installation, in a variety of applications.

Conductors

Conductors are normally made of copper for small to medium size cables and aluminium for larger cables, especially when used in overhead power lines. The reason for this is that aluminium is lighter and is therefore more appropriate, but is not as good a conductor as copper. Carbon has very good conducting properties, which is why it is made into brushes and used in certain types of motors and generators.

The best conducting material is actually silver, followed by copper. Do not be fooled into thinking that gold is a better conductor than copper or silver – it is not. Gold has other useful properties such as it tends to resist corrosion, which is one of the reasons why it is used in spacecraft.

Insulators

Insulation materials used with electrical cables tend to be either thermoplastic or thermosetting, with the development of plastic having largely replaced rubber. A type of thermoplastic called polyvinylchloride (PVC) is used to manufacture twin and earth domestic wiring and has a maximum operating temperature of 70°C but, more importantly, it can be recycled.

Thermosetting, on the other hand, is normally used to manufacture electrical plugs and sockets or cables that require a higher operating temperature range such as 90°C. Butyl is such a cable installed around electric water heaters, its thermosetting insulation offering a higher level of protection from heat damage. Unlike thermoplastic, thermosetting materials cannot be reused.

Fire alarm systems and other potentially hazardous environments such as those found in chemical plants are often installed using mineral insulated cable, sometimes referred to as Pyro. The insulator within this type of cable is called Magnesium Oxide (MI), which enable it to withstand temperatures of up to 1000°C.

Porcelain is also used as an insulator for cable supports on electricity pylons but overhead cables themselves tend not be insulated through any specific material other than air. This is why overhead cables pose a danger to anybody conducting activities nearby. There have been many reported cases of fatalities involving metal ladders, scaffold and elevated work platforms but also, unbelievably, through graphite fishing rods, because their construction includes carbon.

Let us look at a case study of a very tragic story, which highlights such dangers and hopefully will serve as a warning to others.

Craig Gowans was a 17-year-old promising footballer and was on a full-time contract with Falkirk Football Club. Unfortunately, he came into contact with overhead power lines whilst moving 20-foot-long poles used for transportable nets, which are positioned behind football goals to catch balls that miss their target. This tragic story indicates that any activity in and around overhead cable lines is potentially hazardous.

Following Craig's death, Falkirk was fined £4000 for a breach of health and safety regulation but they also carried out a full review of their procedures and implemented many changes. However, such changes are classified as 'reactive' because they occurred after the incident! Had they carried out a review before this tragic event it would have been classified as 'proactive'!

Tackle two difficult words: reactive and proactive

Reactive: reacting to something after the event such as an accident

Proactive: taking action by causing change before an accident happens

Electrical circuits

Figure 2.30 shows a simple lighting circuit, but please remember even the most complicated will also include the same core components. Also shown are some conductors and insulators, with each element performing a particular task.

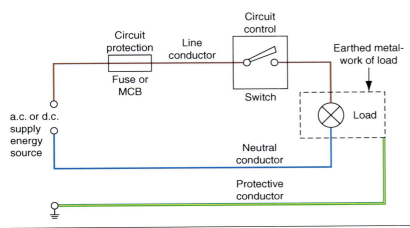

Figure 2.30 All electrical circuits include core components such as a busbar, protective device, switch and an electrical load

- The supply, which could be a battery, generator or the mains, is normally fed on to the bus-bar, which is a strip of metal.
- All the protective devices such as fuses or circuit breakers sit on the bus-bar and their function is to operate and isolate the circuit when a fault occurs.
- Every circuit has to be controlled and this is provided by the switch. When the switch is open then the circuit is incomplete and electricity cannot flow. Closing the switch completes the circuit and electricity is allowed to flow.
- The term load is sometimes used to describe a heavy weight but it also applies to electrical circuits to describe items such as lights, heaters or motors. In other words, the term electrical load is the actual item you want to operate or work.
- There are three conductors being used in the diagram: the line conductor which delivers current to the load, and the neutral, which returns it to the supply. Also shown is a protective conductor, sometimes referred to as the earth conductor, but its proper name is circuit protective conductor (c.p.c).

The c.p.c performs a very important role. During a fault condition, certain metallic elements of an electrical installation could became live, in which case the c.p.c – which forms part of an earthing system – would offer the fault current a nice juicy low resistance path. This in turn will cause the circuit's protective device to operate and isolate the circuit.

Conductor/insulator activity

Put an X in the correct box that identifies whether the materials shown are either a conductor or insulator. One has been done for you.

Put an X in the correct box	Conductor	Insulator
Copper		
Air		
Steel	x	
Iron		
Ceramic		
Paper		
Thermoplastic material		
Gold		
Wood		
Carbon		
Magnesium Oxide		
Aluminium		
Thermosetting material		

Learning outcome 2.7

List the sources of electromotive force.

The term electromotive force is used whenever electricity is generated and, generally, there are three ways of achieving this. The first was described earlier regarding how an electric current is passed through a solution and an electrochemical effect causes negative particles to be collected on one electrode and positive particles on the other. This chemical effect is therefore how a battery builds up an electric charge and is therefore a source of potential energy. The term battery however actually means a collection of cells, for instance a car battery will include eight 1.5 V cells all connected in series. This means that the total amount of electromagnetic force is: 8 x 1.5 V = 12 V. There are generally two types of battery cell, known as primary and secondary. Primary cells can only be used once, but secondary cells can be recharged. Secondary cells are more commonly called rechargeable batteries.

We have already covered the term electromagnetism previously, but an e.m.f is generated whenever an electrical conductor is moved across a magnetic field. This is known as the generator effect. Various methods are used to create the mechanical energy required to move the conductors, such as nuclear, oil and coal, which are all used to generate high volumes of steam. The Dinorwig Power Station in North Wales is a hydro scheme that uses a reservoir of water, which is released deep inside a mountain in order to turn several incredibly large turbines. The water is actually released at peak times when there is a high demand for electricity. Tidal systems and wind farms on the other hand make use of certain forces of nature to create the motion required to generate an e.m.f.

Important point

If an electrical conductor is passed through a magnetic field then electricity is created.

Another means of producing an electromotive force is known as a thermocouple and is an item which uses two dissimilar metals to create a sensor, which, when heated produces an e.m.f. For example, Alumel and Chromel are two dissimilar metals which are combined into a device, called a thermocouple, and are placed around the jet pipe of an aircraft to sense its temperature. The e.m.f produced by the thermocouples is actually direct current, which is then passed through a set of electrical cables into a cockpit indicator. The pilot is then able to monitor this critical temperature.

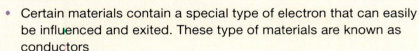

Believe it or not

Batteries, kettles, thermocouples and circuit breakers all operate using two dissimilar metals.

Time out, let's summarise some **key** points!

- Electricity was once thought of as tiny particles that flow from a positive charge to a negative charge. This is logical when you think about it, but there is one major issue involved in this notion – it is wrong
- We now know that electricity is caused by the attraction of negative electrons towards a positive source
- Certain materials contain a special type of electron that can easily be influenced and exited. These type of materials are known as conductors
- Other materials resist electrical current flow and are known as insulators
- Switches are used to control the flow of electricity, because electricity cannot flow if its path is incomplete (switch open), but can flow if its path is complete (switch closed)
- A fuse or circuit breaker is used to protect the cable. For instance if a fault occurs then too much current will flow and will generate a lot of heat that actually melts the cable insulation
- Hopefully, though, this excess amount of heat will either trip or blow the circuit breaker or fuse in order to isolate the problem circuit
- All wind turbines, micro hydro (water wheels) and vehicle alternators actually produce electricity in the same way; the only difference being is the type of energy used to turn a conductor, which is sat inside a magnetic field

To review your understanding, try this True or False quiz based on conductors and insulators.

Question 1: The movement of electrons is what gives us electricity.

True	False

Question 2: Gold is the best conducting material.

True	False

Question 3: Air is used as an insulator.

True	False

Series circuit

Circuit A which is shown at Figure 2.31 is a typical series circuit and there are particular rules which apply to it with respect to voltage, current and resistance. For instance:

- There is only one path for the current to flow and that is through resistor R1
- There is only one load (R1) therefore all the suppply voltage (20 V) will be applied across it

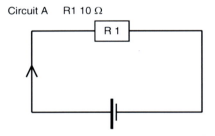

Circuit A   R1 10 Ω

Figure 2.31   **Figure 2.32**

Although the circuit current (I) is not shown in Figure 2.31, it can be calculated by using the Ohm's law triangle (Figure 2.32). Simply put your finger on the letter I to extract the formula.

$$I = \frac{V}{R}$$

$$I = \frac{20}{10}$$

$$I = 2 \text{ amps}$$

The total amount of current that is therefore flowing in circuit A is 2 amps.

In circuit B (Figure 2.33) , the supply voltage (V) is not shown but we can also calculate it by using the Ohm's law triangle. Put your finger on the letter V to extract the formula.

$$V = I \times R$$

$$V = 2 \times 100$$

$$V = 200 \text{ V}$$

Circuit B   R100 Ω

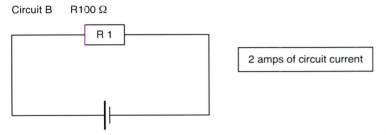

R 1

2 amps of circuit current

Figure 2.33

The supply voltage for this circuit is 200 V.

In circuit C (Figure 2.34), although the value of R1 is not shown we can calculate it by using the Ohm's law triangle.

Put your finger on the letter R to extract the formula.

$$R = \frac{V}{I}$$

$$I = \frac{50}{2}$$

$$R = 25 \ \Omega$$

Important point

The same current flows through all the resistors that are connected in a series circuit.

Circuit C

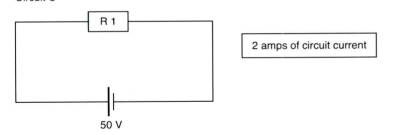

R 1

2 amps of circuit current

50 V

Figure 2.34

The total resistance of the circuit is therefore 25 Ω.

Circuit D (Figure 2.35) is slightly diffferent because this time it contains two resistors in R1 and R2.

Circuit D

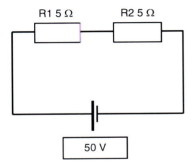

R1 5 Ω   R2 5 Ω

50 V

Figure 2.35

When calculating the total resistance of a series circuit we simply add up all the resistor values.

Total resistancce = R1 + R2

Total resistancce = 5 Ω + 5 Ω

Total resistancce = 10 Ω

Because both resistors are the same value (5 Ω) the supply voltage will be shared equally between them. This means that 25 V will be dropped across both R1 and R2.

With reference to circuit D, we can also prove all this by calculation, again using the Ohm's law triangle.

We known that V = I x R.

We have already established the total resistance: 5 Ω + 5 Ω = 10 Ω.

Then, using the Ohm's law triangle, extract the formula for the circuit current I:

$$I = \frac{V}{R}$$

$$I = \frac{50}{10}$$

I = 5 amps (remember the same current flows through all the components in a series circuit)

Calculate the voltage dropped across R1 and R2:

Volt drop across any resistor = circuit current x resistor value

- Volt drop across R1 = 5 amps x 5 Ω

  = 25 V

- Volt drop across R2 = 5 amps x 5 Ω

  = 25 V

The voltage dropped across each resistor will equal the supply voltage:

V R1 (25 V) + V R2 (25 V) = 50 V (Supply)

Circuit E

Yet again circuit E (Figure 2.36) is a series circuit, but this time the resistors have different values and therefore the following applies: the majority of the suppply voltage will appear across R2, the largest resistor.

To prove this we first need to calcuate the total resistance just like before.

Total resistance = R1 + R2

Total resistance = 20 Ω + 30 Ω

Total resistance = 50 Ω

We then need to calculate the circuit current

$$I = \frac{V}{R}$$

$$I = \frac{100}{50}$$

I = 2 amps

We can also calculate how much voltage is dropped across each resistor

- Volt drop across R1 = 2 amps x 20 Ω

  = 40 V

- Volt drop across R2 = 2 amps x 30 Ω

  = 60 V

V R1 + V R2 = 100 V (Supply)

The majority of the suppply voltage appears across R2.

Important point

To calculate the total resistance of a series circuit add up the resistance values.

Circuit E

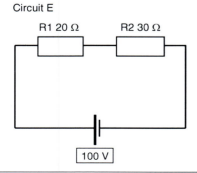

Figure 2.36

Resistors in parallel

The other method of connecting electrical circuits is called parallel and in fact, most electrical circuits are connected in this way.

Circuit F below (Figure 2.37) shows how a parallel circuit is connected and there are particular rules, which apply with regard to voltage, current and resistance. For instance:

- Supply voltage is common to all the loads
- The circuit current will divide and flow down each individual branch
- The largest current will flow through R3 (lowest value resistor)
- The total resistance will be less than the lowest value (<5 Ω)

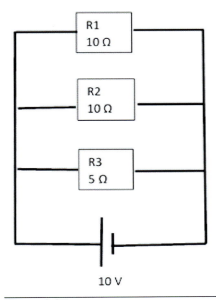

R1
10 Ω

R2
10 Ω

R3
5 Ω

10 V

Figure 2.37

Calculating the total resistance of a parallel circuit is very different to a series circuit.

There are many ways of doing this. Some methods use fractions, another approach only works with two resistors. I always recommend that students use a scientific calculator.

Scientific calculators will include the reciprocal function, which looks like:

1/x

or

x^{-1}

Circuit G

Calculating total resistance is straightforward but if you miss a step then your answer will be completely wrong.

For example, to calculate the total resistance of a parallel network such as circuit G (Figure 2.38), you need to enter all the resistor values in sequence:

10 followed by x^{-1}

+

10 followed by x^{-1}

+

5 followed by x^{-1}

Top tip

The total resistance of a parallel circuit is always less than the smallest value.

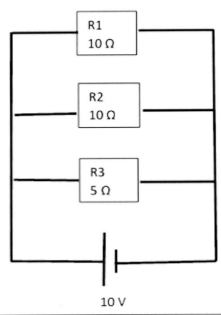

Figure 2.38

Once all the resistor values have been entered in, push the equal sign (=).

Vital step

It is vital that you enter x^{-1} again one last time to finish off.

The answer should be = 2.5 Ω.

Notice that when calculating resistance in a parallel circuit, the total resistance is always lower than the smallest value: 2.5 Ω< 5 Ω.

Once the total resistance of a parallel has been worked out, if you are given the value of the supply voltage, then you can work out the total circuit current.

Again, by using the Ohm's law triangle, we can calculate the total circuit current for circuit G

Figure 2.39

$$I = \frac{V}{R}$$

$$I = \frac{10}{2.5}$$

$$I = 4 \text{ amps}$$

We can also work out the value of the current in each branch. Remember for a parallel circuit, voltage does not change (it is common to all branches).

Current through R1:

$$I = \frac{V}{R1}$$

$$I = \frac{10}{10}$$

$$I = 1 \text{ amp}$$

Current through R2:

$$I = \frac{V}{R2}$$

$$I = \frac{10}{10}$$

$$I = 1 \text{ amp}$$

Current through R3:

$$I = \frac{V}{R3}$$

$$I = \frac{10}{5}$$

$$I = 2 \text{ amps}$$

Total current = IR1 + IR2 + IR3

Total current = 1 + 1 + 2

Total current = 4 amps

Examples

Refer to circuit H (Figure 2.40) and work out the:

- Total resistance of the circuit
- Using the Ohm's law triangle calculate the circuit current

Example

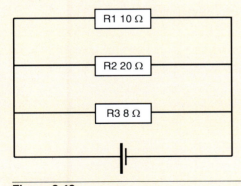

Figure 2.40

Top tip

When using your calculator to work out the total resistance of a parallel circuit:

You must push the = sign after you have entered in all the resistor values.

You must then enter x^{-1} one final time.

Time out, let's summarise some **key** points!

- Series circuits are those that only contain only one path for the circuit current
- Another way of stating this is that current is common to all loads in a series circuit
- To work out the total resistance of a series circuit then we add up all the resistor values. If the resistors in a series circuit are equal in value then the supply voltage is shared between them
- Parallel circuits contain a number of branches, which means that a little bit of the circuit current will flow down each
- The same amount of voltage is applied across each parallel branch
- Another way of stating this is that voltage is common to all loads in a parallel circuit
- To work out the total resistance then we **cannot** simply add up all the resistance values. We have to take the reciprocal value of the resistors involved
- This can be done easily through the x^{-1} or $1/x$ function on your calculator, remembering that you must push equals after all the values have been entered and x^{-1} or $1/x$ must be pressed one last time
- The total resistance of a parallel circuit, will always be less than (<) the value of the smallest resistor

Fill in the gaps

Fill in the gaps below with the following words:

parallel, series, out, on, voltage, current, blows, add, less

In a ___ circuit the ___ is common to all components. To work out the total resistance you simply ___the resistance values. Christmas tree lights used to be wired in series, but one major disadvantage is that if one light bulb___, all the lights go___.

In a ___ circuit the ___is common to all components. A very important rule for parallel circuits is that the total resistance is always ___ than the smallest value. Modern Christmas tree lights tend to be wired in parallel, because if one light bulb blows, the others remain ___.

Learning outcome 2.10

Recognise how electrical instruments are connected.

Electrical test equipment is used to measure different elements of electrical circuits in order to ensure that it meets the requirements of:

- The Wiring Regulations
- Safety systems
- Manufacturer's data

Test equipment is also used when conducting fault finding techniques, trying to trace down faults by comparing the reading to various symptoms with the actual readings given by the test equipment. Although three common measurements – voltage, current and resistance – can be measured through three separate meters known as a voltmeter, ammeter and low resistance ohmmeter they can also be measured through what is commonly called a multi-meter. A typical example is shown in Figure 2.41.

Figure 2.41 A multimeter is able to measure a variety of measurements such as voltage, current and resistance

Let us now examine each individual meter in turn, starting with a **low resistance ohmmeter**. This type of meter is always used on a circuit that has been made dead, that is all forms of electrical energy has been removed from it. A low resistance ohmmeter does not need the supply to operate because it has an internal battery, which will send out a small current. A healthy circuit will permit this current to return to the meter and indicate a low resistance reading (circuit complete).

Establishing continuity of protective conductors is a vital safety feature, because it ensures that a fault current will chose to flow down the protective conductor and not through any person who might come in contact with the circuit. The

Important point

To check if a circuit is complete a low resistance ohmmeter would register a very low resistance such as 0.05–0.08 Ω.

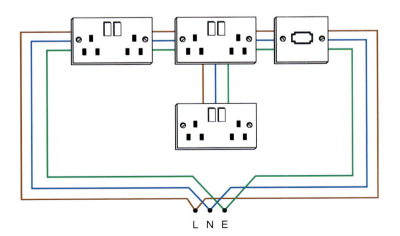

L N E

Figure 2.42 Continuity of all the conductors is vital if a circuit is to operate correctly and safely

Description 🕐

An ammeter is connected in series.

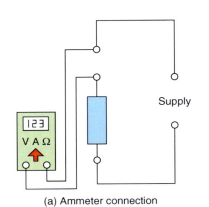

(a) Ammeter connection

Figure 2.43 (a) An ammeter has a very low internal resistance and must always be connected in series

Description 🕐

A voltmeter is connected in parallel.

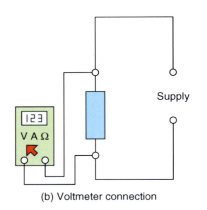

(b) Voltmeter connection

Figure 2.43 (b) A voltmeter has a very high internal resistance and must always be connected in parallel (across a component)

resistance value of a healthy and continuous protective conductor should be very low such as 0.05–0.08 Ω It is also important to remember that the actual leads of an ohmmeter contains some resistance and therefore it is very important to take away the value of the leads from all measurements.

An **ammeter** is used to measure the amount of electrical current that flows in a circuit. To measure current, an ammeter has to be inserted in series so that it becomes part of the circuit as shown in Figure 2.43a. The resistance of an ammeter is therefore designed to be as low as possible to minimise any disruption.

A **voltmeter** unlike an ammeter is connected across a component. The reason for this is that a voltmeter measures the difference in potential between two points, which is why voltage is also called potential difference or even electrical pressure. Measuring across something is referred to as connecting in parallel and is shown in Figure 2.43b. Remember it is like the voltmeter is looking at the circuit from outside, which is why its resistance needs to be as large as possible, so that only a little bit of the circuit is disturbed.

When using a meter as a voltmeter or ammeter there is an expectation regarding safety that the black lead is connected first, followed by the red positive terminal. In certain circumstances, connecting the red lead first could be dangerous. This is because if a person touched a live supply whist holding the black lead, the circuit could be completed through that individual. Disconnection should be in the reverse order the red lead disconnected first followed by the black lead.

Learning outcome 3.1

Identify the different diagrams and drawings used in electrical work.

Figure 2.44 Multimeter leads are normally coloured red (positive) and black (negative)

An electrician will use a range of diagrams and drawings in order to sketch, draw and outline the construction of a building. They are also used to explain how components are related and how they operate or work. They include:

Block diagrams

A block diagram is a very simple diagram in which the main circuit elements are represented by rectangular or square shapes. The purpose of a block diagram is to show how these circuit components relate to each other. A typical example is shown below in Figure 2.45, which reveals all the core components found in every electrical circuit.

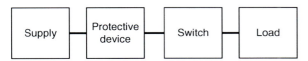

Figure 2.45 A Block diagram is used to simplify and identify the main components involved in a circuit

Circuit diagrams

A circuit diagram is used to show how a circuit works and the main components are represented by graphical symbols. Its main purpose is to help an electrician gain full understanding of what must be done in order for the circuit to work. This is especially useful when conducting fault-finding techniques. The example shown below though Figure 2.46 is a circuit diagram for a single switch controlled lighting circuit. Closing the switch will therefore cause the light to operate.

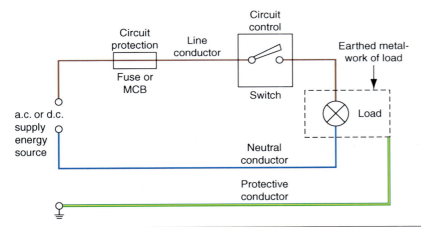

Figure 2.46 A circuit diagram uses graphical symbols to indicate what must be done to operate the circuit

Wiring diagrams

There is an old expression, which started out as a television advertising slogan, which stated 'it does exactly what is says on the tin' meaning you should be able to work out what something does from its description. This applies to a wiring diagram, because it serves to explain how to physically wire up a circuit. This means that all the different connections must be made known so that all the conductors can be connected into the correct termination. An example is shown below in Figure 2.47, which indicates all the interconnections involved regarding a one-way lighting circuit.

Junction box

Blue   Blue

                                        Lamp

Brown   Brown

Neutral   CPC   Line

Brown   Blue   Brown sleeve

C

One-way switch

Figure 2.47 Wiring diagrams indicate precisely which terminations the circuit conductors should be connected to

Learning outcome 3.2

Identify graphical symbols used in diagrams and drawings.

Graphical symbols are used extensively in electrical and electronic engineering in order to make drawings, diagrams and charts understandable to those who use them. All symbols in the Wiring Regulations comply with BS EN 60617, in order to convey the same meaning to a manufacturer, designer or installer. The On-Site Guide is a very handy book in this regard because its back cover contains many of the symbols used in electrical installation. Figure 2.48 shows many of the graphical symbols used in wiring systems.

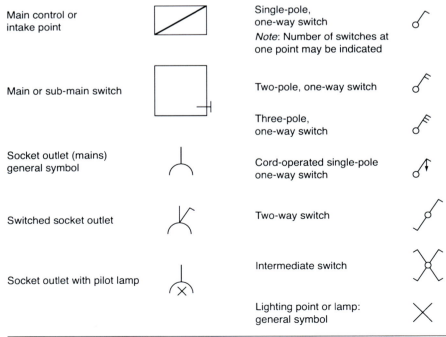

Main control or intake point

Main or sub-main switch

Socket outlet (mains) general symbol

Switched socket outlet

Socket outlet with pilot lamp

Single-pole, one-way switch

Note: Number of switches at one point may be indicated

Two-pole, one-way switch

Three-pole, one-way switch

Cord-operated single-pole one-way switch

Two-way switch

Intermediate switch

Lighting point or lamp: general symbol

Figure 2.48 All symbols in the Wiring Regulations comply with BS EN 60617

Continued

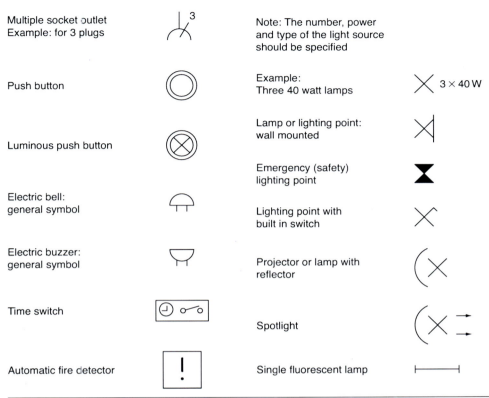

Multiple socket outlet Example: for 3 plugs		Note: The number, power and type of the light source should be specified
Push button		Example: Three 40 watt lamps $3 \times 40\,W$
Luminous push button		Lamp or lighting point: wall mounted
Electric bell: general symbol		Emergency (safety) lighting point
		Lighting point with built in switch
Electric buzzer: general symbol		Projector or lamp with reflector
Time switch		Spotlight
Automatic fire detector		Single fluorescent lamp

Figure 2.48 When symbols are standardised then everyone involved in both design and installation uncerstand what they represent

Learning outcome 3.3

Recognise the typical circuits in a domestic dwelling.

Domestic properties tend to be installed using PVC flat profile cable, better known as twin & earth (Figure 2.50), which is cheap to produce and is mainly installed in existing walls, floors or through wooden roof joists. The main supply to the property will normally be delivered using a PVC/SWA cable (Figure 2.49), chosen because it is often buried underground. Its steel armouring therefore will protect it from any mechanical impact and damage.

Domestic properties use different systems to supply both lighting and power. Lighting circuits are normally split into two separate circuits which feed and control downstairs and upstairs lights, with each circuit being protected

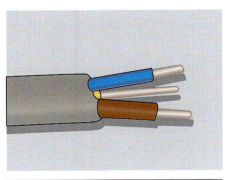

Figure 2.50 Domestic properties are normally insta led with twin & earth cable

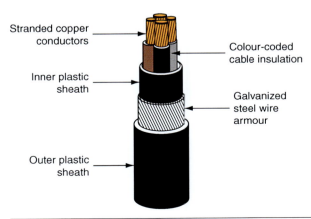

Stranded copper conductors

Colour-coded cable insulation

Inner plastic sheath

Galvanized steel wire armour

Outer plastic sheath

Figure 2.49 SWA cable is very strong and offers a lot of mechanical protection

by a separate protective device and should be labelled as such in the consumer unit.

In domestic environments, lighting circuits are normally wired with what is called a loop-in system or through a junction box. Both incorporate an arrangement whereby a supply is fed into a room and is then fed out to all the other rooms in sequence on that particular floor. Various controls are used, such as one-way lighting control, whereby a single switch controls a sole circuit.

Two-way Switch

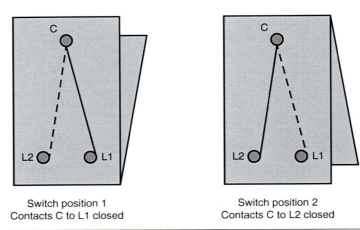

Switch position 1
Contacts C to L1 closed

Switch position 2
Contacts C to L2 closed

Figure 2.51 With two-way switching, either switch can make or break the lighting circuit

Description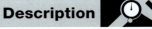

Loop-in is a type of lighting circuit common in domestic properties.

When control is required, such as in a domestic staircase or when lighting is required in a large living room with multiple doors, then a two-way lighting control is used (Fig 2.51). Another system is known as intermediate switching, which introduces several extra switches and is a very useful system if you need to provide lighting in very long corridors or a stairwell that links several floors in a block of flats.

Description

Ring and radial are types of power circuits.

Power sockets are also split into two general types and they are called ring and radial. A radial circuit feeds each socket outlet in a particular circuit but then stops when it arrives at the last one in that particular circuit. A ring circuit on the other hand, once it has fed the last socket is returned to the consumer unit and back to the same protective device. This type of arrangement is shown in Figure 2.52.

There are other differences between a ring and a radial circuit and this is seen in the maximum area that these circuits can supply.

For instance:

- A radial circuit if fitted with a 2.5 mm² cable is limited to 50 m²
- A radial circuit if fitted with a 4.0 mm² cable is limited to 75 m²

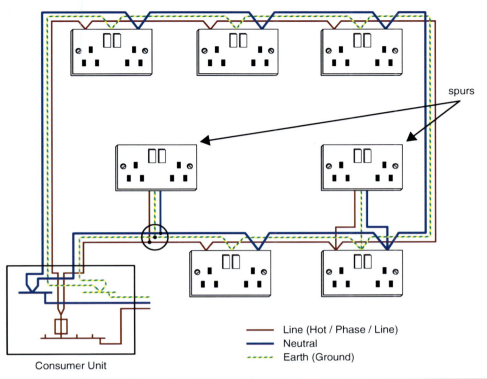

Figure 2.52 Ring circuit conductors are returned to the consumer unit

A ring circuit is normally installed with a 2.5 mm² cable and any number of sockets can be fitted as long as the maximum area it supplies is not bigger than 100 m². An extra unfused single or double socket can be fed from each socket outlet contained in the ring. This is known as spurring. If fused spurs are used then any number of extra sockets can be fitted.

Domestic properties tend to use three different ring circuits, which are then used to supply and feed:

- Downstairs sockets
- Upstairs sockets
- The kitchen

A kitchen needs its own ring circuit because it houses a lot of electrical equipment but especially equipment such as kettles and toasters which are designed to draw a lot of current very quickly.

Electric cookers should not be fed from a ring circuit because they too draw an awful lot of current when operating and would certainly overload it. This is why it is always recommended that equipment such as electric cookers, showers and water heaters larger than 15 litres should be supplied through their own individual (dedicated) circuit. Such circuits should also be controlled through a double pole isolator as indicated through Figure 2.53 – a type of switching arrangement which ensures that both the line and neutral connections are made or broken simultaneously (together).

Important point

Modern domestic properties normally contain three ring circuits.

Figure 2.53 Very large electrical loads such as cookers, showers and water heaters should be controlled by a double pole isolator

Time out, let's summarise some **key** points!

- Voltmeters possess a very high internal resistance and should be connected across components (in parallel) in order to measure a potential difference (volts)
- An ammeter has very low internal resistance and is connected in series in order to measure the amount of electrical current that is flowing in a circuit
- Ohmmeters contain a battery, which passes a current around a circuit. If the circuit is complete and heathy the meter will register a very low reading Ω (Ohm)
- BS EN 60617 is used to standardise the use of symbols across Europe
- Various diagrams are used to show different types of information such as a circuit diagram which is used to indicate how a circuit actually works
- Domestic properties are normally divided into power circuits (radial and ring), lighting circuits as well as dedicated circuits for certain items such as electric cookers, showers and water heaters

To review your understanding try this True or False quiz based on very useful electrical installation knowledge.

Question 1: Voltmeters should be connected in parallel.

True	False

Question 2: A wiring diagram tells you how a circuit operates.

True	False

Question 3: Loop-in and junction box are two methods of wiring domestic lighting.

True	False

Question 4: Any number of sockets can be fitted within a ring circuit.

True	False

Question 5: Modern domestic properties should be installed with two ring circuits.

True	False

Electrical SI and derived units

Match the meaning of the following by placing the appropriate number
next to the letter in the box provided (the first one has been done for you).

| Ampere A | A = 1 | Electrical current (1) |

| Coulomb B | B = | Electrical power (2) |

| Voltage C | C = | Electrical resistance (3) |

| Ohm D | D = | Electrical charge (4) |

| Watt E | E = | Energy (5) |

| Joule F | F = | Potential difference (6) |

Electrical terms and items

Match the meaning of the following by placing the appropriate number
next to the letter in the box provided (the first one has been done for you).

Water
A

A = 2

Item used to step voltage
up and down. Only works
with alternating current
(1)

Direct current
B

B =

Can be compared to an
electrical current
(2)

Mains Voltage
C

C =

Domestic 230 V single
phase supply
(3)

D

D =

Type of electricity used in
batteries
(4)

An electrical cable
can be compared to
E

E =

Ohm's law triangle
E
(5)

Transformer
F

F =

A water pipe
(6)

Electrical principles word search

Find the following hidden words:

CPC, VOLTAGE, POWER, FUSE, PARALLEL, MAGNETSIM, LOAD, MICRO, RING, AMPERE, ALTERNATOR, TRANSFORMER, WATT, CHEMICAL, SWITCH, RADIAL, ELECTRONS, RESISTANCE, NEUTRONS, SERIES, AREA

C	P	C	L	V	I	C	L	I	M	Q	K	P	A	F	U	S	E
A	R	E	A	N	O	N	H	M	O	M	E	O	T	P	M	R	O
S	E	R	I	E	S	L	E	E	T	E	S	W	A	A	S	K	H
R	Q	V	K	M	V	Y	T	G	M	C	D	E	M	R	N	I	M
I	A	C	C	W	O	O	E	A	R	I	L	R	Y	A	M	M	N
N	K	D	O	E	L	C	R	E	G	Q	C	E	E	L	I	S	T
G	A	M	I	S	T	T	A	T	K	E	L	A	O	L	N	I	D
L	M	S	P	A	H	C	T	I	W	S	K	W	L	E	O	T	A
E	G	A	T	L	L	V	O	S	E	R	I	N	G	L	R	E	O
E	L	E	C	T	R	O	N	S	S	F	T	X	E	S	C	N	L
R	E	S	I	S	T	A	N	C	E	D	A	T	A	G	I	G	O
S	S	N	O	R	T	U	E	N	T	I	M	E	A	G	M	A	G
A	L	T	E	R	N	A	T	O	R	G	F	U	L	W	R	M	M
A	M	P	E	R	E	K	R	E	M	R	O	F	S	N	A	R	T

Test your knowledge

1. Know the fundamental units of measurement used in electrotechnical work.

1. Which of the following is an SI base unit?
   a. time
   b. acceleration
   c. volume
   d. area

2. Which of the following is an SI derived unit?
   a. time
   b. length
   c. ampere
   d. acceleration

3. Electrical current is measured in:
   a. volts
   b. ohms
   c. amps
   d. watts

4. Voltage is also known as:
   a. potential difference
   b. power
   c. resistance
   d. amperes

5. Electrical resistance is measured in:
   a. volts
   b. ohms
   c. amps
   d. watts

6. Milli is represented by:
   a. 1/10000, 0.0001
   b. 1/10, 0.1
   c. 1/100, 0.01
   d. 1/1000, 0.001

7. Kilo is represented by:
   a. 10000
   b. 10
   c. 100
   d. 1000

8. 3 kV is equal to:
    a. 300 V
    b. 30,000 V
    c. 30 V
    d. 3000 V

9. 0.03 amps is equal to:
    a. 3 mA
    b. 30 mA
    c. 300 mA
    d. 3000 mA

10. The area of an object is calculated by:
    a. multiplying its length by its width
    b. dividing its length by its height
    c. dividing its length by its width
    d. multiplying its length by its height

11. Volume is measured in:
    a. m°
    b. m^1
    c. m^2
    d. m^3

2. Know the basic principles of electrical circuits.

12. An electric bell operates through a:
    a. heating effect
    b. chemical effect
    c. magnetic effect
    d. power effect

13. Which of the following operates through a chemical effect?
    a. contactor
    b. relay
    c. electric motor
    d. battery

14. What kind of circuit is shown below?
    a. parallel
    b. power
    c. series
    d. serial

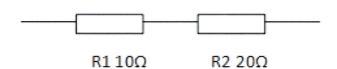

R1 10Ω   R2 20Ω

15. Calculate the total resistance of the circuit below:

    a.   30 Ω

    b.   10 Ω

    c.   6.66 Ω

    d.   200 Ω

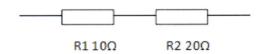

R1 10Ω   R2 20Ω

16. What kind of circuit is shown below?

    a.   parallel

    b.   series

    c.   constant current

    d.   power

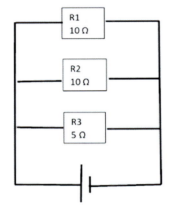

17. Calculate the total resistance of the circuit shown below:

    a.   15 Ω

    b.   2.5 Ω

    c.   5 Ω

    d.   1000 Ω

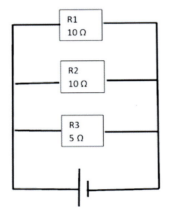

18. Using the Ohm's law triangle, work out which of the following formulas is correct for extracting resistance?

   a. $\dfrac{\text{amperes}}{\text{watts}}$

   b. volts x amperes

   c. watts x amperes

   d. $\dfrac{\text{volts}}{\text{ampere}}$

19. Metre (b) below is measuring?

   a. voltage

   b. power

   c. resistance

   d. current

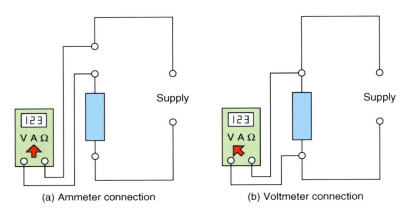

(a) Ammeter connection   (b) Voltmeter connection

20. The meter below is measuring:

   a. voltage

   b. power

   c. resistance

   d. current

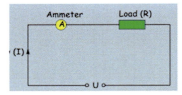

21. If 2 amps of current flows through a 10 Ω resistor, calculate the circuit voltage:

   a. 0.2 V

   b. 20 V

   c. 12 V

   d. 8 V

22. An electric circuit is supplied by a 20 V battery and has 10 amps of current flowing through a single resistor. Calculate the value of the resistor:
     a. 2 Ω
     b. 200 Ω
     c. 0.5 Ω
     d. 30 Ω

23. An electric circuit consists of one 5 Ω resistor and is supplied by a 20 V battery. Calculate the circuit current:
     a. 15 amps
     b. 25 amps
     c. 0.25 amps
     d. 4 amps

24. An electric circuit consists of 20 amps of current and is supplied by a 10 V battery. Calculate the circuit power:
     a. 200 volts
     b. 20 watts
     c. 200 amps
     d. 200 watts

25. An electric circuit consists of one 20 V battery which has developed 100 watts of power across a single resistor. Calculate the circuit current:
     a. 2000 amps
     b. 5 amps
     c. 80 Amps
     d. 0.2 amps

26. If the total current in the circuit is 6 amps and 2 amps flows through A1, how much current is flowing through A2?
     a. 2 amps
     b. 6 amps
     c. 4 amps
     d. 12 amps

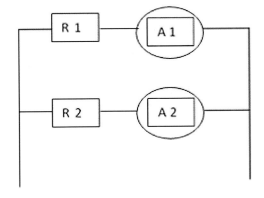

27. If 5 amps flows through ammeter A1, how much current flows through A2?
    a.   10 amps
    b.   5 amps
    c.   2.5 amps
    d.   7.5 amps

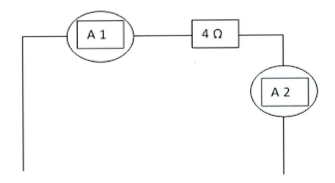

28. If 2 volts is dropped across R1, how much voltage is dropped across R2?
    a.   10 volts
    b.   8 volts
    c.   12 volts
    d.   4 volts

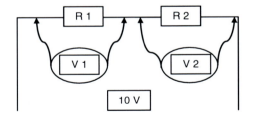

29. If 20 volts is dropped across R1, how much voltage is dropped across R2?
    a.   20 volts
    b.   10 volts
    c.   5 volts
    d.   40 volts

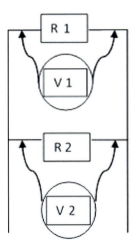

30. Electrical current is created when:
   a. electrons move
   b. protons move
   c. neutrons move
   d. nucleus move

31. Which of the following controls an electrical circuit?
   a. load
   b. switch
   c. bus-bar
   d. fuse

32. A light bulb matches which term?
   a. load
   b. switch
   c. bus-bar
   d. fuse

33. Which of the following protects an electrical circuit?
   a. load
   b. switch
   c. bus-bar
   d. fuse

34. Which of the following is generated by a battery?
   a. e.m.f
   b. resistance
   c. current input
   d. chemical output

35. **TWO** of the following terms are associated with conductors. But which ones?
   a. switch
   b. line
   c. c.p.c
   d. load

36. A domestic mains supply matches which of the following?
   a. direct current
   b. direct power
   c. d.c.
   d. alternating current

37. A battery produces which of the following?
   a. direct current
   b. direct power
   c. alternating power
   d. alternating current

38. **TWO** of the following are conducting materials. But which ones?
    a. copper
    b. ceramic
    c. air
    d. aluminium

39. **TWO** of the following are known as good insulators. But which ones?
    a. gold
    b. plastic
    c. silver
    d. air

40. Which of the following is the best conducting material?
    a. gold
    b. copper
    c. silver
    d. iron

41. Which of the following is the second best conducting material?
    a. gold
    b. copper
    c. aluminium
    d. iron

42. Voltmeters should always be connected:
    a. in series
    b. in parallel
    c. outside the circuit
    d. across the switch

43. Which of the following is used to measure electrical current?
    a. ampmeter
    b. volt meter
    c. ohmmeter
    d. ammeter

44. Which of the following is used to measure the continuity of conductors?
    a. amp meter
    b. high resistance ohmmeter
    c. low resistance ohmmeter
    d. ammeter

3. Know basic aspects of diagrams and circuits.

45. A circuit diagram is used to indicate how a circuit:
    a. works
    b. is wired
    c. is installed
    d. is tested

46. A block diagram indicates the main components involved using:

    a. triangles

    b. circles

    c. squares

    d. oblong

47. What kind of diagram is shown below?

    a. circuit diagram

    b. assembly diagram

    c. wiring diagram

    d. block diagram

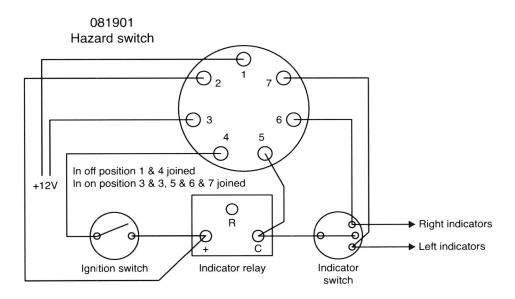

48. What kind of symbol is shown here?

    a. socket outlet

    b. double socket outlet

    c. double switched socket outlet

    d. light switch

49. What kind of socket outlet is shown here?
   a.   1 gang
   b.   2 gang
   c.   un-switched double socket
   d.   double pole socket

50. A typical modern domestic installation will contain how many ring circuits?
   a.   1
   b.   2
   c.   3
   d.   4

51. The typical cabling for use in electrical installations in domestic premises would be:
   a.   metal conduit
   b.   sheathed and insulated PVC thermoplastic
   c.   MICC
   d.   metal tray

52. What kind of lighting circuit would most likely be found in the vicinity of a staircase?
   a.   single
   b.   two-way
   c.   attic method
   d.   intermediate

53. The maximum floor area which a ring circuit can feed is:
   a.   100 sq. metres
   b.   75 sq. metres
   c.   50 sq. metres
   d.   25 sq. metres

54. Power circuits in a domestic property are known as:
   a.   ring and tone
   b.   line and neutral
   c.   radial and circuit
   d.   ring and radial

Chapter 2 checklist

Learning outcome	Assessment criteria – the learner can:	Page number
1. Know the fundamental units of measurement used in electrotechnical work.	1.1 Identify basic (SI) units of measurement for general quantities. 1.2 Outline the SI or derived SI unit for electrical quantities. 1.3 List the common multiples and sub-multiples used w thin electrotechnical work.	50 56 57

2. Know the basic principles of electrical circuits.	2.1 Determine basic electrical quantities using Ohm's law.
	2.2 List the effects of an electric current.
	2.3 Calculate values of electrical power in basic circuits.
	2.4 Outline the fundamental principles of an electrical circuit.
	2.5 Outline the basic principles of direct and alternating current.
	2.6 Identify between materials which are good:
	  • Conductors
	  • Insulators.
	2.7 List the sources of electromotive force.
	2.8 Measure voltage on an electrical circuit.
	2.9 Determine electrical quantities in simple circuits.
	2.10 Recognise how electrical instruments are connected.
		60
		68
		
		65
		
		76
		
		78
		84

3. Know basic aspects of diagrams and circuits.	3.1 Identify the different diagrams and drawings used in electrical work. 3.2 Identify graphical symbols used in diagrams and drawings. 3.3 Recognise the typical circuits in a domestic dwelling.	86 88 89

EAL Unit: QACC1/02

Environmental protection measures

EAL Electrical Installation Work – Level 1. 978-1-138-23206-8

Learning outcomes

The learner will:

1. Know the applications of energy sources used in the building services engineering industry.
2. Know the methods of reducing waste and conserving energy while working in the building services engineering industry.
3. Know how to safely dispose of materials used in the building services engineering industry.
4. Know the methods of conserving and reducing wastage of water within the building services engineering industry.

Learning outcome 1.1

Outline the types of energy used in properties.

Learning outcome 1.2

State the importance of reducing carbon emissions from buildings.

A lot is said about the wonder of modern technology, such as remarking about how super computers have mapped all the genes in the human body; the progress in developing ever more complex brain surgery as well as fighting and finding cures for diseases such as cancer is both astonishing and extraordinary. But have you ever wondered if any of this would have been possible without electricity?

Certainly, in the Western world we take technology for granted alongside mains electricity, gas supplies and a clean and stable water source. We rely on such services to wash ourselves, prepare meals, watch television, surf the internet, charge our mobile devices and keep us warm.

Electricity is actually created by combining two of the four forces of nature: magnetism and electricity. They are interrelated, in other words you cannot have one without the other. If an electrical conductor is moved at right angles across a magnetic field then electricity is generated.

Different materials and even elements of nature are used to create the motion required to turn conductors through this magnetic field and include: water, wind, nuclear and fossil fuels such as coal, oil and gas. Energy from the sun is also used to generate electricity and is known as a photovoltaic (PV) system which acts as a transducer – changing light into electricity.

But is there a price to pay for developing and using all this technology?

The term environment describes the world we live in and relates to our immediate surroundings. Let us now examine all the various technologies involved in producing electricity but also reflect how each impacts on our environment.

Did you know?

There are four forces of nature: gravity, electromagnetic, strong nuclear and weak nuclear!

Did you know?

Microphones and loud speakers are also transducers but which one changes sound waves into electrical signals and which one changes electrical signals into sound? Why not research the answer on the internet.

Tackle a difficult word: transducer

A device that changes one form of energy into another

Did you know?

- Coal was formed through organic plant matter.
- Oil and natural gas was formed through tiny plants and animals being subjected to pressure and heat.
- Fossil fuels are projected to last for only another 75 years.

Key point

When electricity is generated through the use of fossil fuels a lot of greenhouse gases are created, which in turn intensify the effects of global warming.

Fossil fuels

Certain types of fuels are known as 'fossil fuels' because they were formed from the fossilised remains of prehistoric plants and animals that took millions of years to form. Fossil fuels include coal, oil and gas and are non-renewable; in other words, they will not last forever.

Apart from being fossilised, coal, oil and gas have another thing in common in that they are all used to heat up water in order to generate steam.

The various different stages involved in generating electricity are shown in Figure 3.1. Various fuels are used to:

- Provide ignition and combustion in order to create steam
- Steam is then used to drive a turbine
- The turbine drives a generator
- Electricity is then passed through a transformer in order to be transmitted and distributed

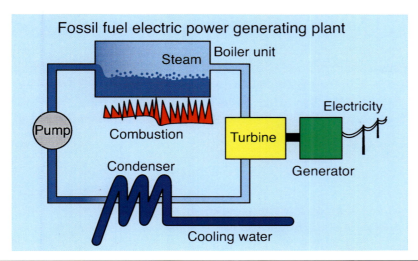

Figure 3.1 Fossil fuels are used to heat water in order to create steam. Steam in turn drives large turbines which are mechanically linked to generators, which create electrcity

There is a price to be paid for using fossil fuels, since they contribute to the production of carbon dioxide. Although carbon dioxide is produced naturally such as when people exhale and when materials decompose, fossil fuels are used so extensively across the world that dangerous amounts of carbon dioxide (CO_2) is produced. This amount of CO_2 produced differs in various sectors as indicated through the pie chart (Figure 3.2).

Carbon dioxide is a greenhouse gas and these types of gases alongside pollution are the main cause of global warming and as well as other environmental factors create the following effects:

Depletion of the ozone layer. The ozone layer is a layer of naturally occurring gas that sits above the surface of the earth and serves as a shield to protect people from harmful ultraviolet radiation produced by the sun. Due to some chemicals found in the release of pollution such as sulphur, the ozone layer is under constant attack and large holes start to appear. It is feared that the extra volume of ultraviolet radiation that will get through will cause higher levels of skin cancer.

ACC02/5

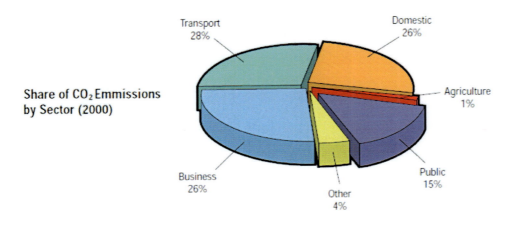

Figure 3.2 Every time we travel, buy food, goods or even clothes we generate carbon dioxide

Important point

Acid rain affects animals and plants which means it affects the balance of nature. This is why our immediate environment is changing.

Believe it or not

At one time even the gas inside aerosols and fire extinguishers contributed to global warming, but the ones in use today use gases that are ozone friendly.

Acid rain. When fossil fuels like coal and oil are burned they produce sulphur and in turn this produces acid gases. A lot of these acid gases are propelled into the sky and combine with clouds. When those clouds produce rain, the actual water content will be acidic or, in other words, acid rain. The problem with acid rain is that it affects plants, trees, animals and fish. It can also contaminate water systems that are used for drinking as well as damaging buildings because it attacks and eats into materials such as glass, plastic and even brick.

Increase in air pollution. Air pollution can be carried over long distances. This means that the acid rain in Norway is caused by air pollution in Britain and other European countries.

Increase in air temperature and increased ocean levels. An increase in air temperature at the South and North Poles has led to vast glaciers breaking off but such occurrences have also caused sea levels to rise, which threaten low-lying countries, such as Holland, which are especially vulnerable to flooding.

Droughts and severe change in weather patterns. Countries tend to get used to their weather patterns, but when they experience unseasonal climatic conditions are thrown off kilter. For instance, in Britain it does not take a lot of snow to fall before it causes chaos on our roads. Compare that to countries such as Norway, Sweden and Canada, who are used to living with severe levels of snow. Other countries are affected by droughts, where many weeks or months elapse before any sight of rainfall, and hurricane and cyclones are common in certain countries. However, because the environment has been under attack over many years, many countries and regions are now experiencing changing weather patterns which experts believe is a direct result of climate change.

Time out to consider some important facts about fossil fuels and global warming:

- Fossil fuels, water, wind and the sun are all used to generate electricity
- Fossil fuels however generate a lot of carbon dioxide which is a greenhouse gas
- This creates problems such as: depletion of the ozone layer, acid rain, air pollution and a rise in air temperature which leads to a rise in sea levels

Because of how certain fossilised fuels affect climate change, several low carbon systems were developed such as:

- Solar thermal
- Solid fuel (biomass)
- Hydrogen fuel cells
- Heat pumps
- Combined heat and power (CHP) and combined cooling, heat and power (CCHP)

We will now consider each in turn.

Solar thermal

This system uses solar panels that act as transducers and change the sun's rays into thermal energy to generate domestic hot water. Figure 3.3 illustrates how the solar panels work alongside a conventional boiler which is used on colder cloudy days. However, some carbon dioxide is still produced.

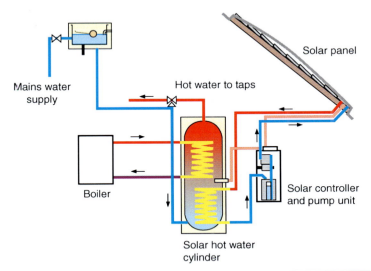

Figure 3.3 Solar thermal acts as a transducer, changing the sun's rays to heat, which is then used to produce hot water

Biomass

Biomass uses living, or recently living organisms or materials – for example, wood products such as pellets, dried vegetation and plants and even animal by-products – to produce energy. Another type of biomass is converted into liquid fuels and this is called biofuels. The burning process involved generates low levels of carbon dioxide although it is claimed that this can be taken up quickly by the growing of new plants.

> **? Did you know?**
>
> A camp fire is an example of biomass producing energy (heat).

Hydrogen fuel cells

Water is a compound because it is a substance which consists of atoms from two or more different kind of elements. This is why its chemical symbol is H_2O – every one of its molecules contains two atoms of hydrogen and one of oxygen. Water, of course, is the most abundant compound on our planet and, in fact, the human body is made from about 60% water (some of us a little more). Of these two atoms, hydrogen has already been used to generate energy but unfortunately in a very destructive way when hydrogen bombs were developed and used against two Japanese cities: Hiroshima and Nagasaki in World War II.

Nuclear fission is when the atom is split, but these days, thankfully, much more positive use is being developed and made of hydrogen through vehicle fuel cells. Cars are either driven through what is termed an internal combustion engine – petrol or diesel – or, alternatively, a large collection of batteries is used to power electrical motors (electric cars). A hydrogen cell is similar to a normal battery and has similar components: a positive and negative electrode as well as an electrolyte which acts to separate both electrodes.

> **? Did you know?**
>
> Space rockets are powered with hydrogen power cells.

Hydrogen gas is used and stored in a special fireproof tank which is required due to the fact that hydrogen is potentially explosive. The gas is fed to the positive terminal and oxygen is fed to the negative electrode. A chemical reaction occurs which puts hydrogen protons on the positive terminal and electrons on the negative terminal. This process generates what is termed a potential difference, or a voltage.

These types of cells are pollution free and the only limiting issue is the amount of hydrogen gas in the tank. As fossil fuels are depleted, more and more will hopefully be spent in developing this type of fuel source.

Heat pumps

There are two popular types of heat pumps which are used as an alternative energy source and, although much reduced, both types still contribute to the levels of carbon dioxide produced. Heat pumps extract heat (energy) from the air or ground which is then used to heat a property. For instance, a lot of energy from the sun is actually retained in the ground, therefore a ground source heat pump system is designed to extract this heat by transferring it to a circulating fluid contained in polythene pipes. The actual pipes are buried in the ground, in purpose-built trenches as shown in Figure 3.4. A heat exchanger is then used to extract the energy or heat from the fluid which is then used to provide underfloor heating and hot water.

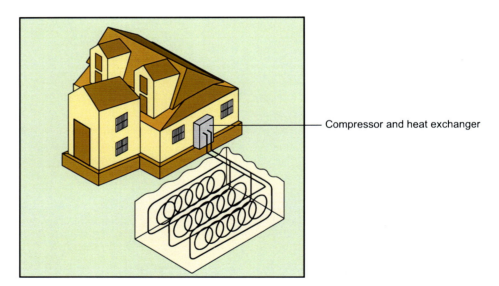

Compressor and heat exchanger

Figure 3.4 Ground source heat pumps extract solar energy that has been stored in the ground

Air source heat pumps work in a similar fashion but extracts heat from the air. It might be useful to think of it as the opposite to a refrigerator. A fridge extracts heat by using a coolant, which leaves the inside of the fridge cold so that our food stays fresh for longer and reduces it from being infected with bacteria.

The air source heat pump therefore uses a refrigerant system of evaporation involving a compressor and a condenser to absorb heat in one place and then release it in another. There are three distinct stages which are shown in Figure 3.5. An evaporator changes the liquid or refrigerant into a gas and a compressor then causes its temperature to rise considerably. A heat exchanger is then used to extract the heat and this is used to heat up our homes. The

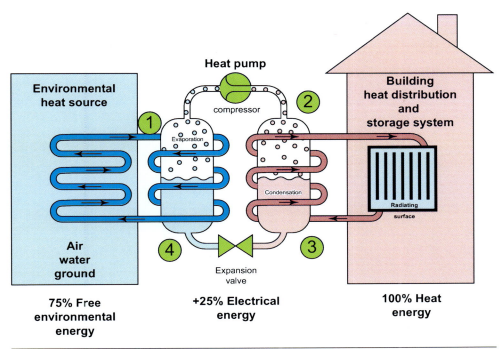

Figure 3.5 Air source heat pumps acts like a fridge in reverse, extracting the heat from air, which is then used to heat our homes

last stage is a condenser which changes the gas back to a liquid as well as an expansion valve which is used to make this process more efficient. The cycle is then continued and repeated.

Combined heat and power (CHP)

Combined heat and power (CHP) is a system that combines the production of both usable heat and power (electricity) in a single highly efficient process known as cogeneration. Historically, the methods used to create electricity generated a lot of heat that was wasted and was often seen, and still is in certain places, as large clouds of steam rising from high cooling towers. Capturing this useful heat creates a system that is 80% efficient in comparison to 50% when the steam was deliberately allowed to escape.

This sort of wastage is also associated with conventional domestic gas boilers, but a Micro CHP system recovers and redirects this heat which is then used to feed a small heat engine which then drives a generator to produce electricity. This makes this system very efficient since it makes far more use of the energy involved; it also reduces the amount of carbon dioxide produced. One major disadvantage with CHP systems is that there is less need to produce heating during the summer months, which means the opportunity to create additional electricity is limited.

A further system is known as tri-generation since it produces and combines elements of cooling, heat and power (CCHP), but even though this is a highly efficient system it still produces carbon dioxide.

Zero carbon technology

At the beginning of this chapter we described how electricity is produced by the use of fossil fuels such as oil, coal and gas which are all used to create steam, which is then used to mechanically drive a turbine in order to generate electricity.

 Description

Going Green refers to the processes and technology that are renewable and friendly to the natural environment.

? Did you know?

A generator accepts mechanical power in order to create electricity.

A motor does the complete opposite: accepts electrical energy in order to produce mechanical energy.

The mechanical energy required to generate electricity can also be obtained by harnessing elements of nature therefore we will look at each in turn.

Wind power

Wind power (Figure 3.6), also referred to as micro-turbo, uses and directs air flow to drive a turbine and is seen as an alternative power source. Despite the fact that no carbon dioxide is produced, there are objections to the installation of huge wind farms, on the basis that they spoil areas of outstanding natural beauty as well as being quite noisy for those who live nearby.

Figure 3.6 Wind turbines are classified as a form of renewable energy

Wave energy

Description

Renewable is defined as those resources that are naturally replenished, such as solar energy, rain and wind.

Waves are created when the energy from the wind is transferred into huge swells and surges in the sea. Waves then cause the water levels to rise and fall which can then drive specially designed and located turbines in order to generate electricity. Wave energy is potentially one of the most environmentally friendly forms of electricity generation as it is clean and renewable, and it has huge potential across the world. That said, wave patterns are not always predictable and reliable and it is not much use for landlocked countries (countries that have no access to the sea).

Hydroelectric

Did you know?

The Dinorwig Power Station is renewable but not green.

The Dinorwig Power Station is a hydroelectric scheme built inside a mountain – Elidir Mawr. It is known as a pumped storage system because it operates by transferring water between a reservoir located above the mountain (Marchlyn), down through several huge turbines located inside the mountain and then finally down to Lake Peris below. This is why it is known as the Electric Mountain. The water is stored in Marchlyn in order to anticipate high demand on the National Grid. For instance, when popular television programmes are scheduled or even sports events such as the Six Nations or Wales participating in the European Football championship semi-finals, then the amount of electricity required can be very substantial indeed. It is during these peak times that Dinorwig Power Station comes online in order to generate enough electricity to boost the National Grid.

The water is then pumped back from Llyn Peris and to Marchlyn during off-peak times or what are known as quiet hours (after midnight). Whilst Dinorwig Power Station is renewable, since it operates through the use of water, it is not green. This is because it actually uses more electricity when pumping the water back up than it generates. However, the money gained by selling the electricity at peak times, in comparison to buying it during off-peak hours, means the scheme paid for itself in 10 years.

If the water supply is constant, then the amount of electricity generated would also be consistent and these sort of schemes are known as run-of-river schemes. A further system is possible, known as a micro-hydro scheme, whereby small streams or even rivers can drive their own turbines through what is basically a water wheel. Lastly, flowing water can also generate electricity that is contained in large storage system made up of huge batteries. Known as a battery storage power station, the aim is to store electricity that will again be made available during high demand or peak times.

Future developments include the Swansea Bay Tidal Lagoon project, which aims to harness the power within tidal surges in the sea (Figure 3.7) in order to drive specialised turbines to generate electricity. The Swansea Bay lagoon could be the first of six installed around Britain's coastline, which potentially could generate 8% of the UK's electricity.

Tackle a difficult word: commodities

Materials or goods that can be bought and sold, such as copper or coffee

Figure 3.7 Tidal lagoons harness the ebb and flow of at least two tides that occur every day

Learning outcome 1.3

Give examples of how building services engineering industries are working to reduce carbon emissions from buildings.

A Building management system (BMS) is a computer-based system installed in buildings that controls and monitors the building's mechanical and electrical equipment such as:

- Ventilation
- Lighting
- Power supplies
- Fire systems
- Security systems

This means that lighting and heating for unoccupied rooms can be switched off. This type of control system is known as building services engineering and not only is it beneficial since it reduces the amount of carbon dioxide produced, it makes the building far more efficient and cheaper to run. To give you an example, when I was attending a weekend course at a university recently, the course coordinator apologised when none of the rooms used that morning had any heat. The coordinator asked for forgiveness by saying 'I still cannot get my head around the fact, that I have to book all the rooms but also book some heat.' Although security staff had opened all the rooms in preparation for all the different lecturers that day, he had forgotten to inform the building services department to heat rooms that were normally unoccupied.

Learning outcome 1.4

Outline the basic operating principles of installations containing environmental energy sources: solar thermal, wind turbine and solar photovoltaic.

Solar photovoltaic is a very popular, carbon-free means of generating electricity by harnessing energy from the sun's rays. Solar cells are typically positioned on roof buildings as shown in Figure 3.8, although they can also be positioned elsewhere. As already stated, solar cells are known as transducers, in other words devices that change one form of energy into another, in this case, light into electricity.

Solar photovoltaic, or solar PV as it is popularly known, changes solar energy into direct current, which means the system therefore requires an inverter to change the direct current produced to alternating current which is in keeping with the National Grid. Solar PV systems actually generate more electricity than the average user needs and therefore the householder can be credited with any electricity that is fed back to the National Grid. The government, despite trying to encourage people to go 'green', has reduced the tariff used to calculate how much the householder would be paid. Thankfully, however, the installation costs of solar PV have also reduced, therefore it is still possible, although over a longer time, to recoup the cost involved. At the time of writing to install a 4 kWh system would on average cost £6,500.

Solar thermal is similar to solar PV, in that solar energy is converted but this time is used to heat domestic hot water. This kind of system can deliver about 1,000 kWh per year which is equivalent to half the amount required and generated through a conventional system. This type of supply would be considered seasonal, producing most of its energy during the summer months, therefore it is often linked to a conventional boiler during the winter months.

Wind generators can either be small or large scale developments and are often called micro-turbo systems. Wind power is a free resource and is far safer than nuclear and does not pollute the atmosphere like fossil fuels. They also require converters so that the direct current generated is able to be used within the home. Wind turbines, however, do split opinion on their use, with many people protesting that they spoil areas of outstanding natural beauty. This is why more and more wind turbines are now being located out at sea. There are also small-scale mini-wind turbines available on the market and these are commonly installed in domestic properties, as indicated in Figure 3.9.

Figure 3.8 Solar panels installed in the Northern Hemisphere should be south facing

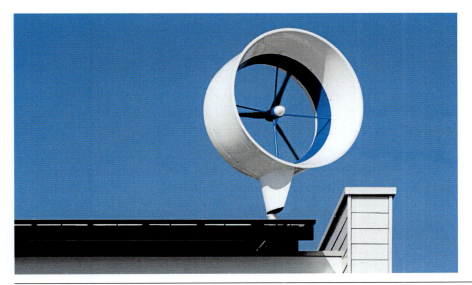

Figure 3.9 Mini-wind systems are smaller scale versions of industrial micro-turbo systems

Carbon classification activity

Put an X in the correct box that identifies its carbon classification. One has been done for you.

	High carbon	Low carbon	Zero carbon	Generates electricity by producing steam
Solar photovoltaic				
Natural gas				
Hydroelectric				
Coal	X			X
Solar thermal				
Biomass				
Wind				

Time out to consider some important facts about low carbon systems:

- Solar thermal uses the sun's rays to heat up water
- Biomass uses living, or recently living, organisms to create energy
- Wave energy harnesses the power of the sea
- Hydroelectric refers to the production of electricity through falling or flowing water
- Hydrogen is used to create a chemical reaction to generate an e.m.f (voltage)
- Air source heat pumps acts like a refrigerator in reverse
- Ground source heat pumps install underground pipes and extract heat from the ground
- Combined heat and power (CHP) combines the production of both usable heat and electricity

Environmental technology activity.

This activity contains two sets of matching terms and each matching pair should be given the same number. One has been done for you: 'Solar PV = 2' matches 'System that uses the sun to produce electricity = 2'

Solar thermal	Solar PV		System that uses the sun to produce heat	System that uses the sun to produce electricity
= 1	= 2		=	= 2
Biomass	Hydrogen fuel cell		Fuel is made by combining hydrogen and oxygen	Material derived from living, or recently living organisms.
=	=		= 3	= 4
Air source heat pump	Ground source heat pump		Uses pipes installed in the ground	Opposite to a fridge, extracts heat from the air outside
= 5	= 6		=	=
CHP	CCHP		Fuel in; heat and electricity out	System that combines elements of cooling, heat and power
=	=		= 7	= 8

To review your understanding further try this environmental technology True or False quiz.

Question 1: Wind turbines do not contribute to global warming because they do not generate any carbon dioxide.

True	False

Question 2: Solar PV can be thought of as a transducer because it changes one form of energy into another.

True	False

Question 3: A hydroelectric system (running water) is renewable.

True	False

Learning outcome 2.1

Outline the working practices that can be employed to conserve energy and protect the environment.

Learning outcome 2.2

Indicate the methods used for reducing material wastage.

The term 'environment' describes the world in which we live, work and play; it relates to our immediate surroundings.

All the material that we use in the construction industry is known as the 'built environment'. When we produce materials not only do we use natural resources but we also produce carbon dioxide which then adds to global warming. It is important, therefore, to consider how we can reduce our impact on the environment, both within the construction industry and as individuals.

There are many steps that electrical contractors can do to adopt a 'best practice' approach to reducing the amount of waste produced on site. This means taking care when designing a building, including making sure that the materials used are appropriate and fit for purpose. Sometimes cheaper materials are selected but this can lead to premature failure and ultimately have to be replaced.

Being accurate when taking measurements when installing systems will also ensure that no more materials than necessary are used to complete the installation. Reducing waste has to be beneficial not only in reducing the cost to an employer but it can also have a positive effect in reducing the impact on the environment.

Contractors can also store more materials than required and in some cases ever need, instead of planning each project carefully and only ordering what

? Did you know?

The environment describes our natural world, the world that surrounds us.

Tackle a difficult word: built environment

Material that we use to build houses and other buildings

is actually required. Whilst insufficient stock leading to delays is never ideal, managing stock levels in a way that does not permit excess waste allowance is far better for costs and reduces the waste stream. Modern day contractors also increasingly use what are termed modular systems, which are factory installed and tested assemblies This means that all the contractor has to do is physically fit them in place. This is why they are sometimes referred to as plug and play.

Current government incentives encourage the view that waste should be reduced if possible but, if this is not achievable, reusing it. Consequently, a waste policy looks to protect human health and therefore the impact on the environment. This is known as a hierarchy of waste management (Figure 3.10), a five-step process with minimising and preventing waste as the preferred option.

Reuse

Certain elements of leftover materials can be fit for purpose for future use. This means short lengths of cable, reusable leftover lengths of conduit, trunking or trays should be returned to the store area for safe storage and future use. It also means that the costing involved in any further jobs will be reduced, which increase a company's profit margin.

Recycle

Recycling actually involves everyone in all walks of life, including those who work and live in domestic, commercial and industrial environments. Recycling means that items are used again, which in turn means that it reduces the requirement for such items to be freshly produced. Recycling reduces the impact on the actual environment, because it reduces the amount of carbon dioxide generated.

Energy capture

This phase involves capturing energy when waste is burnt, that would otherwise be lost.

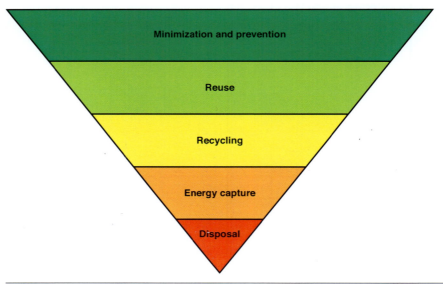

Figure 3.10 Whilst minimising and preventing waste is the preferred option, landfill should be seen as the last option

Disposal

Waste sent to landfill or incineration without recovering any amount of energy

Collective responsibility

We all have a collective responsibility when it comes to the environment. This is not just to our families or employer but also to protect future generations through a shift in mindset. However, reducing how we impact the environment does have other benefits because it can reduce what we pay in bills each month.

Measures that people can incorporate include:

- Fitting thermostatic control in order to turn down room or working temperature
- Fitting energy efficient replacement lights such as compact fluorescent light bulbs or LED fittings
- Installing thermal insulation
- Fitting light motion sensors in lighting systems or switching off unused rooms, corridors and working areas

It is also true to say that most people who pay for bills will go around switching lights off and most people who do not pay bills are less concerned and tend to leave them on. Apart from being pro-active when it comes to lights, a great number of people have installed loft insulation, cavity wall insulation and thermostats to reduce the amount of energy they use, as indicated in Figure 3.11. Some properties even include their own sources of renewable energy such as mini-wind, as shown previously in Figure 3.9, or micro-hydro systems, having made use of a stream located nearby.

Figure 3.11 Even small changes such as turning down domestic heating can have a positive effect on the environment and also reduce your bills

The government has also stopped the production of certain wattages of General Lighting Service (GLS) lamps. GLS lamps produce light through a heater made from tungsten, but are not very efficient. Energy-saving light bulbs such as Compact Fluorescent lamps or LEDs are preferred and, although they tend to be more expensive, they are far more efficient and last a lot longer.

Further developments have stemmed from Part L of the Building Regulations, which have introduced a requirement that 25% of all the lights in newly built homes or domestic buildings must be of a low energy type. Furthermore, Energy Performance Certificates have been introduced, which grade the environmental efficiency of a building from A to G, from peak efficiency to the least efficient.

Important point

Sometimes hoping that people buy energy efficient lamps instead of cheap inefficient ones does not work. This is why the government stopped manufacturing GLS lamps and insisted that 25% of all lighting points in new houses must be of an energy efficient type.

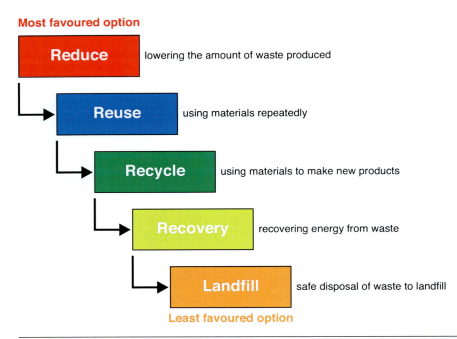

Figure 3.12 Using a 5 step process to reduce our collective effect on the environment requires people to engage with a change in mind-set

Learning outcome 3.1

Indicate the methods of safely disposing of waste material.

Learning outcome 3.2

List the types of materials that can be recycled.

Disposal

It is inevitable that rubbish will be produced on a construction site, but it must be disposed of through a properly devised process. Most packaging material other than cardboard should be skipped, along with all the off-cuts of conduit, trunking and tray that cannot be reused. It is worth remembering that thermoplastic material can be reused, but thermosetting cannot. General waste and site debris will probably go into the same skip and a registered waste carrier company will dispose of the skip contents within a designated local council landfill area or waste disposal site. Only registered waste disposal companies will be given licences for this type of work and will require a waste transfer note. Remember, sending material to a landfill site is the least favoured option. As shown in Fig 3.12, landfill is a last resort.

Typical materials that can be recycled include: paper including newspaper, plastic bottles, cans, glass bottles, oil, metal, wood, batteries, mattresses and garden cuttings. White goods or electrical goods such as microwaves, fridges, and toasters are controlled through what is known as WEEE – Waste Electrical and Electronic Equipment recycling.

Some materials are classified as hazardous waste – items such as fluorescent tubes, batteries, oils and especially asbestos. A consignment note is required to highlight the presence of hazards.

Learning outcome 3.3

Identify what action to take if work activities endanger the environment.

Laws have been produced to protect how people and organisations impact on our environment and companies adopt what is called environmental management systems to ensure that they comply with these requirements. The Environmental Protection Act 1990 was a very important ruling of law and those who do not abide with it will be prosecuted. Environmental Protection Officers regularly look at how companies dispose of their waste by taking samples of water supplies to ensure that they are not contaminated.

Any individual or company that suspects that they have caused an incident or accident which might have affected the environment must report this to the relevant authority. The consequences of not reporting such issues could have far serious implications, especially when considering exposure to asbestos since it is cumulative – every exposure builds upon the previous ones. All employees should report such instances to their immediate superior or the Environmental Protection Officer.

Learning outcome 4.1

State the importance of water conservation within buildings.

Learning outcome 4.2

List the methods for reducing water wastage.

Learning outcome 4.3

Indicate the methods available for capturing surface and the uses of water and recycling used water.

When it comes to energy saving measures then this includes saving on water and this is certainly applicable to bathrooms, which makes up almost 60% of domestic water use. There are many simple ways of achieving this and we will now examine them in turn.

Flow reducing valves. Pressure is used in many systems: oil, hydraulics and water flow. The actual pressure is necessary to ensure that the water flows out of the tap at a rate that is favourable to the person, who for instance is either washing or needing water to drink. Too much pressure, however, can cause excess splashing of water, which in turn is actually water wasted. Using spray taps instead of conventional ones can save up to 80% of tap water reserves, although they would not be appropriate if you needed to fill up a sink.

Low volume flush toilets are designed to use a minimum amount of water to operate, which again can lead to a major saving in the amount of water used.

Important point

Reducing water pressure reduces water waste.

Regular maintenance can also be effective, not only in the replacement or tightening of any mechanical fittings and moving parts such as float valves, but also in lagging external water pipes from freezing and repairing or closing off leaky taps.

A positive outlook on the environment also extends to conserving water and this can extend to fitting or installing **water meters** or even smart **gas/electricity meters**. These sorts of measures are encouraged by the government because such schemes are more helpful in getting the user to be responsible for their energy use. This has two positive effects:

- It lessens the effects on the environment
- It can also lead to smaller energy bills

People, through changes in attitudes, can make small changes in their lives that equate to big impacts elsewhere. Washing is a necessary function of our lives and is necessary to keep ourselves clean and healthy. Taking a shower or a bath is normally a personal preference, but a shower is far better and healthier for the environment and can in most cases be carried out in 5 minutes.

Not only will you save money but you can reduce the amount of carbon dioxide produced by 70–80%.

Accept a challenge: can you shower in five minutes?

Why not click on the website below and work out your personal carbon footprint. Every time we drive a car, go shopping or fly in an aircraft we add to our carbon footprint because we are using up resources. Calculate your environmental footprint at http://footprint.wwf.org.uk/

The Energy Saving Trust is an organisation that helps individuals reduce their personal footprint by offering advice on a range of issues including the use of home appliances and installing thermostatic controls, to name just a few. The Carbon Trust, on the other hand, advises businesses how to operate in a sustainable low carbon world, including developing low carbon technologies and solutions.

Learning outcome 4.3.2

Describe the methods rainwater can be collected and stored, underlining the volume of water that could be harvested from common dwellings.

Rainwater recycling

I was born and bred in North Wales where, similar to other parts of the UK – such as Scotland, Northern Ireland and the Lake District – heavy rainfall is part of daily life. Given the high levels of rainfall, and especially that this natural resource is free, more and more people are utilising rainwater harvesting, which is a system that is used to collect and store rainwater by locating specially designed water barrels to catch water that runs off roofs or guttering. Figure 3.13 shows such an arrangement. In domestic situations, any rainwater collected can be used to water your garden, wash your car and even for drinking if fed through appropriate filters. It is also common on farms, when it is used to give animals a clean supply of water.

Figure 3.13 Rainwater harvesting can both water your garden and clean your car for free

Grey water recycling

Grey water recycling is reusing tap water which has already been used once and, rather than flushing it down the drain, is made available again to reduce the amount of water required. Grey water recycling works by taking water that has already been used in showers, baths or even washing machines and after feeding and routing it through appropriate filters, makes it available to operate toilets.

Modern housing and large scale developments, such as constructing a modern school, will look to include rainwater and grey water harvesting as well as solar PV within their actual design. This means that such buildings are self-sufficient and therefore more sustainable and environmentally friendly. However, when buildings or sewage plants do harvest rainwater or treated grey water the machinery involved, especially the use of electrical pumps, will increase greenhouse gas emissions when compared to systems that use mains water.

One last thought about pollution, global warming and how people can undergo a change of attitude which will reduce our personal impact, especially in contemplating how we are destroying our planet: 'If trees could provide free Wi-Fi there would be jungles but sadly they only provide oxygen which we require to live.'

Description

Grey water recycling redirects water that has already been used.

Time out to review what was covered in the last passage and especially how we can reduce our impact on the environment:

- A building management system (BMS) is a computer-based control system installed in buildings to manage and reduce how much energy we use
- Waste should be reduced if possible, but, if this is not achievable, reuse it
- Waste can also be reduced by taking accurate measurements when installing electrical systems
- Recycle materials when possible
- Fitting thermostatic control in order to turn down room or working temperature and energy efficient replacement items such as compact fluorescent light bulbs will reduce our impact
- Flow reducing valves and low volume flush toilets will reduce how much water we use
- Rainwater recycling uses water barrels to catch water that runs off roofs or guttering
- Grey water recycling reuses tap water and is made available to operate toilets

Reuse or recycle activity

This activity contains two sets of matching terms and each matching pair should be given the same number. One has been done for you: 'Reusable left-over lengths of conduit can be = 1' matches with 'reused = 1'

Reusable left-over lengths of conduit can be = 1	Plastic bottles = 2		can be recycled =	reused = 1
Landfill is a =	Hazardous waste =		Fluorescent tubes and batteries are classified as = 3	Last resort = 4
Rain collected to water your garden and washing your car = 5	Grey water recycling = 6		Reusing tap water and recycling it to flush toilets =	Rainwater recycling =

To review your understanding try this True or False quiz based on how we can change the way we live.

Question 1: Thermoplastic material can be recycled.

True	False

Question 2: Recycling helps the environment because fewer items have to be made.

True	False

Question 3: If everyone made small changes in their lives this could help create a big meaningful impact on the environment.

True	False

Microgeneration technologies

Match the meaning of the following by placing the appropriate number next to the letter in the box provided (the first one has been done for you).

A system that installs pipes in the ground and extracts heat from the soil A	A =2	Micro-hydro (1)
A system that uses solar energy to create hot water B	B =	Ground source heat pump (2)
Fuel in, heat and electricity out C	C =	Solar PV (3)
System that changes solar energy to electricity D	D =	Solar thermal (4)
A system that uses free flowing water to create electricity E	E =	Combined heat and power (5)

Let's help the environment

Match the meaning of the following by placing the appropriate number next to the letter in the box provided (the first one has been done for you).

| Landfill
A | A =3 | Carbon dioxide
(1) |

| Every time we buy goods, travel by car, bus, train or aeroplane we generate ...
B | B = | Types of equipment classified as hazardous waste
(2) |

| Fluorescent tubes and batteries
C | C = | Where materials that cannot be reused or recycled are sent to
(3) |

| Built environment
D | D = | Carbon footprint
(4) |

| The total amount of greenhouse gases produced by a person's activities including travelling and the food that they consume is known as their:
E | E = | Material that we use to build houses and other buildings
(5) |

Test your knowledge

1. Know the applications of energy sources used in the building services engineering industry.

1. Which **TWO** of the following types of energy are known to emit high levels of carbon emissions?
   a. natural gas
   b. solar thermal
   c. coal
   d. solar PV

2. Which **TWO** of the following types of energy are known to emit low levels of carbon emissions?
   a. natural gas
   b. solar thermal
   c. heat pumps
   d. coal

3. Which **TWO** of the following types of energy are known to emit zero level of carbon emissions?
   a. wind power
   b. solar thermal
   c. tidal
   d. coal

4. The total amount of greenhouse gases produced by a person is known as their personal:
   a. green house effect
   b. global warming
   c. carbon footprint
   d. depleted ozone layer

5. A solar photovoltaic system converts the sun's energy in order to produce:
   a. gas
   b. electricity
   c. hot water
   d. heat

6. A solar thermal system converts the sun's energy in order to produce:
   a. gas
   b. electricity
   c. water
   d. heat

7. Which **THREE** of the following is used to generate steam?
   a. gas
   b. photovoltaic
   c. coal
   d. nuclear

8. Which of the following would be very suitable to land near windy coastal regions?
   a. tidal
   b. solar PV
   c. micro-hydro
   d. micro-turbine

9. Which of the following is a means of generating electricity from living or recently living organisms?
   a. rainwater harvesting
   b. solar water heating
   c. grey water recycling
   d. biomass

10. Any person who has a stream flowing across their land, could use which of the following to generate electricity?
    a. micro-hydro
    b. tidal
    c. microwave
    d. photovoltaic

2. Know the methods of reducing waste and conserving energy while working in the building services engineering industry.

11. Which **TWO** of the following can reduce the impact on the environment?
    a. travelling in separate cars
    b. sharing a car
    c. travelling on public transport
    d. travelling by taxi

12. If a company or individual causes an incident that has possibly harmed the environment they:
    a. should ignore it
    b. should put up a sign
    c. should hide any evidence
    d. should report it

13. Which three of the following can reduce your domestic bills?
    a. installing thermal insulation in the attic
    b. never switching electrical equipment off
    c. fitting energy saving light bulbs
    d. having a shower instead of a bath

3. Know how to safely dispose of materials used in the building services engineering industry.

14. Which of the following can carry waste?
    a. licensed waste carrier
    b. licensed landfill carrier
    c. registered waste carrier
    d. registered waste contractor

15. Identify the hazardous materials below:
    a. used glass bottles
    b. discarded fluorescent tubes
    c. dead batteries
    d. offcuts of plasic trunking

16. Identify the recyclable materials below:
    a. used glass bottles
    b. discarded fluorescent tubes
    c. dead batteries
    d. offcuts of thermoplastic material

4. Know the methods of conserving and reducing wastage of water within the building services engineering industry.

17. Which **TWO** of the following can reduce water wastage?
    a. spray taps
    b. flushing a toilet twice
    c. using a bath instead of a shower
    d. flow reducing valves

18. Which **TWO** of the following can reduce wastage?
    a   carefully planning the installation stage
    b. cutting more conduit than is required
    c. reusing any left over materials
    d. inacurate measuremens

19. Which three of the following are associated with grey water recycling
    a. bath
    b. showers
    c. washing machine
    d. drinking water

Chapter 3 checklist

Learning outcome	Assessment criteria – the learner can:	Page number
1. Know the applications of energy sources used in the building services engineering industry.	1.1 Outline the types of energy used in properties. 1.2 State the importance of reducing carbon emissions from buildings. 1.3 Give examples of how building services engineering industries are working to reduce carbon emissions from buildings. 1.4 Outline the basic operating principles of installations containing environmental energy sources: solar thermal, wind turbine and solar photovoltaic. 1.5 List organisations which give guidance and advice on energy saving and conservation techniques.	110 110 118 118
2. Know the methods of reducing waste and conserving energy while working in the building services engineering industry.	2.1 Outline the working practices that can be employed to conserve energy and protect the environment. 2.2 Indicate the methods used for reducing material wastage.	121 121
3. Know how to safely dispose of materials used in the building services engineering industry.	3.1 Indicate the methods of safely disposing of waste materials. 3.2 List the types of materials that can be recycled. 3.3 Identify what action to take if work activities endanger the environment.	124 124 125
4. Know the methods of conserving and reducing wastage of water within the building services engineering industry.	4.1 State the importance of water conservation within buildings. 4.2 List the methods for reducing water wastage. 4.3 Indicate the methods available for capturing surface and the uses of water and recycling used water. 4.3.2 Describe the methods rainwater can be collected and stored, underlining the volume of water that could be harvested from common dwellings	125 125 125 126

EAL Unit: ELEC1/05A

Electrical installation methods, procedures and requirements

Learning outcomes

The learner will:

1. Know the key tools and fixings used in electrical installation.
2. Know the basic requirements of wiring support systems.
3. Know the requirements of working with others in electrical installation.

EAL Electrical Installation Work – Level 1. 978-1-138-23206-8

Learning outcome 1.1

Identify the key hand tools and their uses.

Learning outcome 1.2

Identify maintenance requirements for hand tools.

An electrician will use a range of different tools whilst carrying out electrical installation work and these tools will consist of equipment used to mark out, install, terminate and test a wiring system. We will therefore go on a journey and look at the range of tools and equipment that are found and used today.

When onsite, certain checks should be carried well before an installation is even started and includes a pre-work survey to ensure that there is no pre-existing damage to the fabric of the building. A risk assessment should also be carried out in order to highlight any hazards present but equally important an examination of all the tools to ensure that they are both appropriate to the task and fit for purpose. There is nothing more embarrassing than telling a customer that the reason you have 'downed tools' is that you need to disappear for a while in order to fetch some more. It is equally important to examine tools after use, because it allows time to replace or exchange any defective item before it creates delays in any future jobs.

One word of caution: leaving tools scattered around after a job has been completed would not impress a customer or another tradesperson come to that, because this is not what is expected from a paid professional. Safety can also be affected if tools are carelessly left in certain locations that could cause injury to other people and even possibly make contact with live supplies, which is potentially far more hazardous. Drawing from experience, one type of apparatus that are very easily overlooked are test equipment accessories such as insulated probes. This is because, when testing an electrical system, they are actually clipped on to the circuit itself, but are easily forgotten as the cable is withdrawn.

'Test equipment' is associated with two general types of labels. The first is classified as a safety sticker which states that the piece of equipment has been electrically PAT tested for safety but not necessarily in its use. The second is called a calibration label, which states that the equipment has been subjected to an assessment to ensure that its measurements are accurate. Both these labels should specify when they were tested or calibrated as well as indicating when they require retesting or recalibration. Please note that a valid date on either of these labels does not mean that they are safe to use. A full user check should be carried out to ensure that the equipment is physically undamaged and safe to handle.

Safety also applies to hand tools, which in similar fashion should be checked before and after use to ensure they also remain fit for purpose. This includes ensuring that they are cleaned on a regular basis and especially inspecting metal parts when often a light covering of oil is necessary to protect against corrosion.

Stages of installation

Having established that all your tools are in good working order and are fit for purpose, the first task normally involves measuring and marking out exactly where the wiring system is to be installed. On large jobs an electrician will be provided with a layout drawing to indicate the positioning of equipment including the use of a relevant

Top tip

Always carry out a risk assessment to look for any potential hazards before you start work.

Tackle a difficult word: accessories

Items that can be added to something else

Description

PAT stands for Portable Appliance Testing, which means that all electrical equipment is subjected to a safety test. However the user must carry out a pre-use check before they use the equipment to ensure it is physically safe to use.

Important point

Wiring systems need to be appropriate to their environment.

scale so that measurements on paper can be used and transferred to the worksite. The actual specification of the job must be consulted since it is this document that identifies what kind of wiring system is to be installed. For instance, a wide range of environmental factors can affect what type of wiring system is chosen such as water ingress, vibration, corrosion, temperature and possibly impact damage. Therefore, each and every aspect will have to be given due consideration by the person designing the electrical installation. The actual electrical installation, therefore, must be matched to its environmental conditions, which is why various system are used, from sheathed and insulated thermoplastic cables to galvanised metal enclosures.

The actual installation can be marked up either through chalk, marker pen or even a carpenter's pencil, indicating where any equipment will be positioned as well as outlining cable routes as they pass from one room to another. This is especially important when an installation passes from floor to floor, because certain precautions must be in place to stop the spread of fire.

The importance of taking accurate measurements and planning carefully how any wiring system is to be installed is emphasised by a useful saying: measure twice and cut once. This is because inaccurate measurements will inevitably lead to having to redo certain elements, which adds to the cost of the installation as well as affecting the actual work schedule. Unnecessary replacement of materials is also incredibly bad for the environment because you are effectively using more materials than necessary.

Once installed it is important that any electrical system is aesthetically pleasing to the eye, especially regarding the wishes and expectations of the customer. One sure way of displeasing a customer is to install anything twisted or crooked. It could also be referred to as not being 'plumb' which is an expression often used in the construction industry to establish if certain surfaces are level or not as the case may be. This is a very important consideration especially when you are marking up two sides of a room. Taking a measurement from an uneven floor, for instance, will only create further inaccuracies. Far better to look for a datum or reference point, such as an existing prominent feature, in order to take a leading measurement and then use a spirit level to mark out the remainder.

Tackle a difficult word: aesthetic

Pleasing to the eye

Figure 4.1 Spirit levels are very useful to ensure that work surfaces or aspects of electrical installation are parallel and even

Various types of equipment are used during this phase and include tape measures and spirit levels (Figure 4.1) which are used extensively by many different trades but are only good for work involving short measurements.

When measuring long vertical drops a plumb line would be a better option. Early versions were made from lead which is why it's given its name, but modern ones are made from steel. The sheer weight of the pointer tightens the line and indicates true level on a vertical plane.

When you are required to measure across long horizontal walls, then a very convenient piece of equipment is known as a chalk line, which, when made taut, will snap in place and record a true outline made from chalk.

There are modern devices which make finding levels much easier and more accurate. Laser levels fire lasers to establish room area as well as certain points along vertical and horizontal planes. A laser level will effectively self-level, but does require the user to carry out a pre-use check to ensure that the device remains in calibration.

Apart from general marking out equipment, Figure 4.2 highlights that an electrician's tool kit will also include:

- Side snips to cut and prepare cable ends as well as removing any outer sheath
- Screwdrivers to help tighten terminations
- Cable strippers to strip the inner insulation
- Long and short nose pliers to help shape and feed conductors through enclosures

Top tip

A chalk line is very useful when marking out long vertical walls and ceilings.

Top tip

It is always a good idea to etch your name on tools, to stop them walking off to someone else's tool box.

Figure 4.2 An electrician's tools should be insulated along the majority of their length

Important point

When possible, an electrician's tools should be insulated across the majority of its length.

Given the nature of electricity, certain tools are designed specifically with an electrician in mind. An electrician's screwdriver for instance, will be insulated along the majority of its length, again as shown in Figure 4.2. Having marked out the installation, on certain occasions walls have to be chased out and this can be achieved by either a chase cutter or the old fashioned way – using a bolster hammer and chisel. The term first fix is commonly applied to the fitment of various accessories such as metal back boxes, junction boxes and consumer units. Most of these types of accessories are held in place with screws therefore

appropriate screwdrivers are needed to engage various screw types. Once the accessories are in place, second fix signifies the clipping of sheathed cables to a particular surface or drawing unsheathed cables into conduit or trunking. It's important to remember that all wiring systems require appropriate support. PVC flat profile cables, for example, are supported by PVC clips, which are secured in place using an appropriate hammer. Similarly, SWA cables are supported through the use of cleats and conduit can be fitted with a range of different saddles dependant on the surface it is mounted on and the environment it is contained in. Irrespective of the cable or wiring enclosure used, the recommended distances between these supports can be found in Appendix D of the On-Site Guide, a small companion document to the Wiring Regulations.

No prizes for guessing that the next installation stage is known as third fix and refers to the actual termination of the cables. The first task is to prepare the cable end itself and this involves removing both the outer sheath and inner insulation in order to expose the copper conductor. Once this is done, the termination can be inserted and locked in place. An electrician's knife, wire-strippers, snips and screwdrivers are normally adequate tooling to complete this stage in most cases. All electrical connections need to be both mechanically and electrically sound, but over-tightening any connection can severely weaken it. This is why other ways of terminating connections were developed such as crimping. This is a process that uses specialised equipment but is far more precise, since the crimps themselves are matched to a crimping device, often through a colour code. When the termination is crimped, the actual process is controlled by the mechanics of the device and is moulded and shaped down to an optimum depth. This kind of system is known as precision terminating tooling (PTT).

As stated previously, it is incredibly important to ensure that electrical connections are neither under-tightened nor over-tightened, with both circumstances liable to cause high resistance joints and in certain instances electrical fires (Figure 4.3). Torque wrenches or torque screwdrivers are designed to apply an optimal amount of pressure when tightening screws and bolts. The actual definition of the term torque is turning force, such as when a nut is tightened with a spanner. These types of tools can be set to a specific torque setting, such as pounds/inches, and this type of technical detail can be found within manufacturer's information or even engineering maintenance procedures. A good example of this kind of process is found in automotive garages, when a car wheel is reattached following a tyre change. Although mechanics generally use a pneumatic air drill to drive home the locking nuts, the final process involves the use of a torque wrench, which has been set to the required manufactures torque requirements as shown in Figure 4.4.

Case study

This is a personal case study regarding the actions of a former colleague who was carrying out maintenance activities on an RAF helicopter. He was a very experienced individual and actually held the rank of sergeant and was engaged in securing several bolts that were located around the top of the rotor head. When you consider how fast the rotor head of a helicopter turns, then hopefully you can appreciate the importance of this job and the extreme forces involved. In securing the very first bolt he actually sheared it off, but rather than stopping and considering the situation, convinced he had uncovered a serious condition, he tried to tighten the remaining bolts – shearing them all in the process. If this situation was damaging, it was about to get a whole lot worse, because he then

 Description

Second fix relates to clipping the actual wiring system in place.

 **Tackle a difficult word: optimum**

An ideal amount

Key safety point

All electrical connections need to be both electrically and mechanically sound.

Figure 4.3 Badly formed electrical terminations can cause electrical fires

 **Tackle an expression: torque loading**

Applying the correct amount of force to lock an item in place

Figure 4.4 The final part of changing a tyre involves using a torque wrench which has been set to an optimum amount of force to tighten the wheel nuts

tried the other two helicopters that were parked nearby, again shearing all the bolts concerned. Finally, he decided to take stock of the situation and checked what torque loading was stated in the maintenance procedure. To his horror, he realised that instead of pounds/feet, the procedure read pounds/inches. Thankfully most of the helicopters were away that day, otherwise the whole squadron would have been grounded. Experience does help in most situations, but you should never rely on your memory as he did. Always check the procedure.

Termination

Various elements of an installation will require termination ranging from protective devices, switches, loads and circuit conductors. This is why different terminations were developed such as strip connectors, pillar terminal, claw washer, screw head and a nut and washer method as indicated in Figure 4.5.

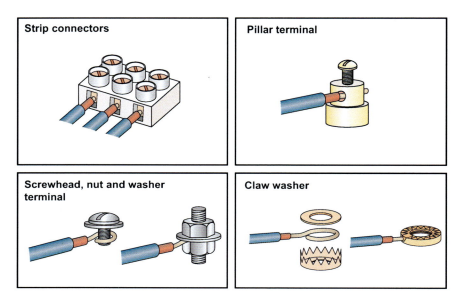

Figure 4.5 Although there are a variety of different types of terminations, their contact resistance should always be as low as possible

Other methods used include soldering, wire wrapping, grub screw and crimping as discussed previously. There is also a method called insulation displacement found in electronics and telecommunications. In general terminations have to be:

- Durable (hard wearing)
- Accessible (reachable)
- Appropriate (suitable to the amount of electrical power used and temperature experienced)

In general, all terminations need to be accessible, so that a detailed examination can be carried out to ensure that the conductors are not broken, exposed or corroded, although soldered connections, given that they are often shrouded or located inside various enclosures, are one exception. There is one characteristic, however, that all terminations share, in that the actual electrical connection must comprise of a very low resistance joint, recording values of between 0.05 – 0.08 Ω. As stated previously, badly formed terminations cause a high resistance joint and, when this occurs, it is one of the main causes of electrical fires.

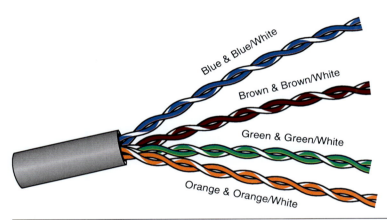

Blue & Blue/White

Brown & Brown/White

Green & Green/White

Orange & Orange/White

Figure 4.6 Data cables are very small and can be quite fiddly to terminate, but damaging their insulation or conductors can introduce interference or even stop your phone from working

 Believe it or not

When electrical fires occur, the actual circuit protective device such as a fuse or circuit breaker might not operate. This is because the overall resistance has actually increased through the badly formed connection.

Because of the importance of ensuring that electrical connections are formed correctly, this is why whenever possible, it is always good practice to double over cable ends. This effectively increases the cross-sectional area of the cable, which in turn lowers the contact resistance.

Learning outcome 1.3

Recognise power tools and accessories.

There are several types of power tools used in the electrical industry ranging for example, as shown in Figure 4.7, from hand drills, impact drivers, nail guns, circular saws and sanders to name just a few. Although all hand-held equipment on a construction site should be powered by a 110 V supply, the safest option is always a battery operated device or even a pneumatic tool, which is powered and operated by pressurised air. It is also worth mentioning that equipment can be powered through either liquid fuel (gas) or hydraulics (oil).

A variety of wiring systems are used in electrical installation and they are fixed to different surfaces and an assortment of materials, so the use of tooling must be both appropriate but also up to the task. For instance, when drilling through stone the job is made much easier if the drill includes a hammer action function,

? **Did you know?**

Pressurised air, if it enters your blood stream, can be fatal because it can cause an air bubble known as an air embolism.

Figure 4.7 A wide variety of power tools are used on site, but they all must be registered, inspected and tested before use

which engages an additional piston inside the drill to increase the actual torque or drilling power. Various types of twist drills are manufactured to meet different applications and dependant on their use their cylindrical shaft and tips are matched to drilling through specific materials as seen below:

- Soft low-carbon steel bits should only be used on wood
- High speed steel (HSS) drills are used when drilling through metal
- Cobalt is used on materials that HSS cannot cut, such as stainless steel
- Hole saws are used to generate large holes in wood or through sheet metal
- A spade bit is used to bore holes through wooden joists
- Masonry bits are designed for use on stonework

There are various ways of terminating an electrical cable and for the majority of situations a selection of insulated, flat blade (Figure 4.8), Philips and Pozidrive screwdrivers can tackle all the different screw heads encountered, bearing in mind what was previously stated about torque setting.

Figure 4.8 A flat blade is only one type of screw driver that an electrician requires

The number of tools an electrician requires therefore can be considerable, which means a good toolbox (Figure 4.9) is essential to house and control tools at the

Figure 4.9 In an ideal tool box, everything has a place, a place for everything

point of work. This, in turn, creates a far safer working environment, since it is less likely that tools will be lost or left unattended whereby they could cause injury or actual damage to the tools themselves. Tool boxes also make the journey to your next job that much easier.

Tools activity

This activity contains two sets of matching terms and each matching pair should be given the same number. The first one has been done for you: 'Pencil = 2' matches 'Item used for marking out wooden products = 2'

Tape measure = 1	Pencil = 2		Item used for marking out wooden products = 2	Used to measure something out =
Hammer =	Screwdriver =		Used to tag cables in place = 3	Used to tighten terminations = 4
Wire strippers = 5	Spirit level = 6		Used to check for level =	Used to strip insulation =
HSS =	Spade drill bit =		Drill bit used on metal = 7	Drill bit used to drill holes through wooden joists = 8

Learning outcome 1.4

Identify types of fixings.

The term 'fixing' is given to specialist anchoring supports used to secure or fix things in place. For instance, putting up a shelf normally involves using the support brackets to mark out where the holes are going to be drilled, then after drilling, a fixing is inserted in to the hole. Screws are inserted through the bracket and into the fixing itself. However, you must ask yourself a very important question: What effect will the shelf and everything being put on it have on the building structure?

Known as load bearing, you must consider how much weight is involved and is the fixing appropriate and fit for purpose. With this in mind, different types of fixings have been developed for a variety of structures and situations.

Plastic plugs

Used for brick, block or tiles, plastic plugs (Figure 4.10) are made from hollow plastic tubes that have been purposely designed for use with a specific size drill.

Description

All fixings are designed to be used alongside a specific size of drill bit and a specific type and size of screw.

Description

Plastic plugs should only be used with brick, block or tiles.

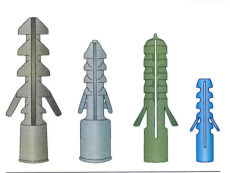

Figure 4.10 Fixings come in all shapes and sizes

Description

Rawlbolt is a type of expansion bolt suitable when securing an installation in stonework.

Description

Spring toggles have been designed for use with hollow walls such as plasterboard partitions.

Using an undersized or oversized drill will both lead to problems, because either the plastic plug will not physically fit through the hole, or it will rotate as you attempt to screw something in place. Fitting an appropriate size screw is equally as important.

Expansion bolt

Rawlbolt (Figure 4.11) is one of the leading manufacturers of fixings which are a type of expansion bolt that have been specifically designed for masonry. Having used an appropriate size drill bit, the expansion bolt is inserted into the hole and tightening the bolt allows a steel ferule to secure it in place.

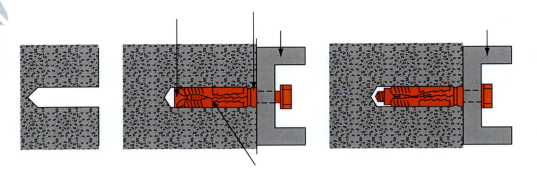

Figure 4.11 Rawlbolt is a manufacture of expansion bolts fixings, which are designed to support loads secured in masonary

Spring toggle

Partition walls are mostly made by using plasterboard and unlike brick or masonry are effectively hollow walls. A special type of fixing, as shown in Figure 4.12, has been designed when installation systems are required in such materials. It is known as a spring toggle or butterfly fixing. Once it has been inserted into the hole, it opens up like a butterfly and then adds some much-needed structural support.

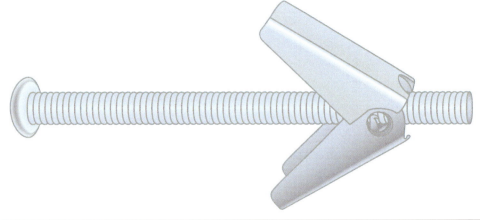

Figure 4.12 Spring toggles have been designed to support loads secured in hollow walls such as plasterboard partitions

Screws and bolts are also used to secure metal conduit, trunking and tray work (Figure 4.13a), but can sometimes work themselves loose, especially in environments exposed to and subjected to vibration. But certain methods can be adopted to secure them in place by various locking devices. For instance, when nuts and bolts are used they should be accompanied by two other elements: a washer and a spring washer. The washer is used to protect the surface that the bolt is impacting on and the spring washer should be placed directly underneath the nut, which then bites into it and stops it working loose. Tab washers, on the other hand, are used to bend over and around a nut, preventing it from rotating loose.

Further examples include a locking device known as a nyloc nut, which contains an inner layer of nylon that resists against the nut if it tries to turn and work its way free. A cable tie (Figure 4.13b), sometimes referred to as a tie-wrap, is a type of fastener used to bind a cable together either when it forms part of a bundle or loom, or when it is secured to a fixed point and hooked through a saddle.

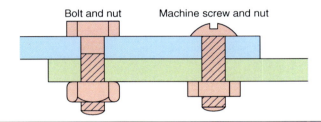

Figure 4.13a

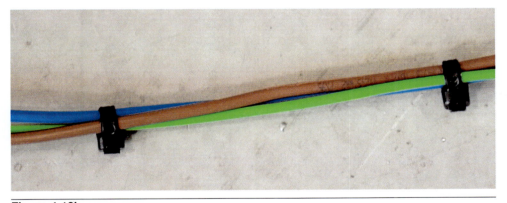

Figure 4.13b

Case study

Tools, equipment and fixings are designed for a particular purpose, but if they are used in the wrong context then sometimes tragedy happens. Kyle Druce was a popular young man from Anglesey, who liked nothing better than to work on his own car. On this one occasion, he needed to work underneath the car and it was raised using two different types of jacks. Jacks are only designed to lift a car for short periods in order to change a wheel; axle stands are far safer and more appropriate to carry out this kind of work. Unfortunately one of the jacks slipped and Kyle was killed. Kyle's father now visits schools and colleges pleading and urging students to always follow health and safety guidelines and always think before they act. There is nothing more powerful than a parent speaking openly about the death of their child, but in reliving this pain, he hopes that future generations will learn from such a tragedy and in doing so has ensured that Kyle did not die in vain!

Fixings activity

This activity contains two sets of matching terms and each matching pair should be given the same number. The first one has been done for you: 'Plastic plug = 2' matches with 'Fixing used with brick = 2'

Plastic plug = 2	Rawlbolt =		Fixing used with brick = 2	Fixing used with masonry = 1
Spring toggle =	Gravity toggle =		Fixing used with plasterboard = 3	Fixing used with long vertical walls = 4

Time out, let's summarise some **key** points!
- Tools form a very important part of electrical work, therefore it is good practice to check tools and equipment before and after every use, this will allow time to replace any faulty or damaged items
- Various types of drill bits are used, matched against different materials such as a HSS drill bit which is appropriate when drilling through sheet metal
- Various types of fixings are used to support equipment such as a spring toggle which is used on hollow walls

To review your understanding, try this True or False quiz which is based on tooling.

Question 1: A chalk line is a type of marking out device that is appropriate when measuring long horizontal surfaces.

True	False

Question 2: A masonry drill bit would be an appropriate choice when drilling holes through a wooden joist.

True	False

Question 3: A rawlbolt type fixing is appropriate when securing equipment on to hollow walls.

True	False

Identify the components of wiring systems.

Recognise the application of wiring system components.

When a wiring system is selected or designed, a great many different aspects will need to be considered. What is required is a wiring system that can meet the needs of the customer but is also safe and suitable to its environment. Certain information needs to be established such as:

- What kind of temperature does the installation have to tolerate including considering any solar radiation that it has to withstand from the sun?
- What kind of impact damage is possible such as: damaging winds, machinery or even rats – because rats have to chew or gnaw at something to stop their front teeth from growing?
- Will the installation come into contact with plants and other organic matter?
- When installing outdoors will corrosion be a problem?
- Does any equipment need to be waterproof?
- Is the environment involved potentially explosive, such as that found within a petrol station or even a flour mill?

Bearing in mind what is listed above we will now go on an electrical journey around various environments and examine why certain systems are selected and identify a variety of components and accessories used in electrical installation.

A journey around a domestic property

Our journey begins at what is termed the 'service head', where the incoming supply is terminated.

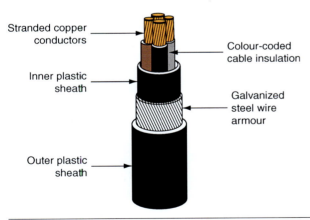

Figure 4.14 When a SWA cable provides the main supply to a building, it is often called a sub main

The incoming supply is normally delivered through a PVC/SWA cable, often referred to as a sub-main. It is selected specifically for this purpose because of its steel armouring which is highlighted in Figure 4.14. SWA cables can therefore provide a lot of mechanical protection which is especially useful since these

Description

The incoming supply in a domestic property is terminated at the service head.

Description

The main service fuse for a modern domestic property should be rated at 100 A.

types of supplies are often buried underground. The sub-main is terminated at what is known as the service head, whose arrangement is shown in Figure 4.15. The equipment involved includes a main service fuse (100 A) which is designed to operate and isolate the supply if any fault develops up to and including the main switch located in the consumer unit.

After the service head the incoming supply continues through two large grey coloured cables which are known as meter tails. When fitted in modern properties, they should be 25 mm² in size, especially if the property contains an electric cooker and/or an electric shower. The tails are then fed to a kWh meter, which the distribution company uses to record and charge how much electricity has being used. Finally, the system also includes an isolator, which is a built-in safety feature because it can be used to isolate the supply, without having to remove the service fuse.

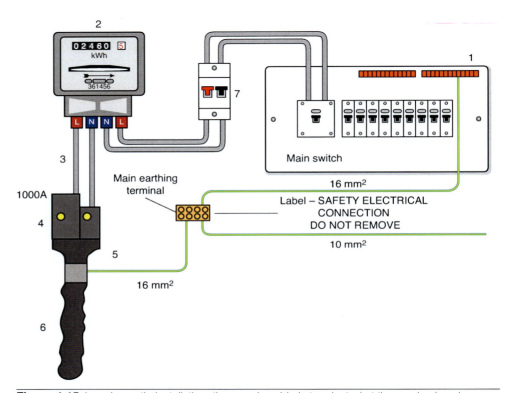

Figure 4.15 In a domestic installation, the supply cable is terminated at the service head

All the equipment involved in bringing power to the consumer unit is the responsibility of the supply company, including the service fuse. This means that even a qualified electrician must gain permission from the supply company to work on any of the service head equipment.

Please note that it is now a requirement since the introduction of Amendment 3 to the Wiring Regulations (BS 7671) that all consumer units installed in domestic properties must be made from non-combustible materials such as metal.

All the electrical loads are split over two distinct circuits, which is why this kind of arrangement is referred to as a split load board and is designed to minimize inconvenience, so that no single fault will affect all circuits. There are certain good practices used when arranging loads; for instance, downstairs lights and upstairs sockets should be ganged together on one side and upstairs lights and downstairs sockets on the other side. Should a fault affect one half of the consumer unit and, for instance, stop the upstairs lights from working, because

Description

17th edition consumer units are known as split load boards and are designed so that no single fault will affect all circuits.

the upstairs sockets are not affected, some degree of lighting can be achieved by plugging in a wall lamp.

All electrical circuits are also given what is termed additional protection, whereby within the consumer unit at least two residual current devices (RCDs) must be fitted. Within domestic properties, RCD's are normally rated at 30 mA and is a device which will operate very quickly by sensing that an electrical current is not being returned through the neutral conductor and being fed down the earthing system. It will also trip under electric shock conditions because it senses that the circuit current is travelling through the person. An RCD, such as that shown in Figure 4.16, is designed to protect people.

Figure 4.16 An RCD protects people, by isolating the circuit, if electrical current is not returned through the neutral conductor and flows to earth instead

When consumer units do not have built-in RCDs and additional protection is required for a specific circuit, a type of circuit breaker with its own built-in RCD called an residual-current circuit breaker (RCBO) can be fitted instead.

In addition to RCD protection, each circuit is also protected by its own individual protective device and this device can be a fuse, circuit breaker or RCBO, each being recognised by a unique British Standard number.

Fuses are designed to be the weakest part of a circuit, which means when an exceptionally large amount of current flows this produces an excess amount of heat and effectively melts the fusing wire. In doing so the fuse has isolated the supply. Three conditions cause excessive currents to flow in a circuit.

1. The first is an overload situation which is when too much current is asked of an otherwise healthy circuit such as overloading mains adapters or extension leads.

2. The second and third are very similar and occur when the line conductor comes into direct contact with the neutral bypassing the load. This is known as a prospective short circuit current.

3. If a similar condition occurs but this time the line conductor makes direct contact with the earthing system – again bypassing the load – then it is known as a prospective earth fault current.

Problems occur, however, if the fuse or circuit breaker fails to operate, then due to the excessive amount of heat produced, the conductor inner insulation will start to melt. This allows bare conductors to come into contact with each other and electrical fires can occur thereafter. Protective devices such as circuit breakers and fuses are fitted in order to protect the circuit cables by isolating the supply.

Description

An RCD provides additional protection against electric shock. However, an RCD only operates on earth faults.

Description

A fuse operates when the heat produced by a fault current or overload melts the fusing wire inside. It is purposely designed to be the weakest part of a circuit.

Description

An overload is when too much is asked of an otherwise healthy circuit. Please note that an RCD will not operate during an overload condition.

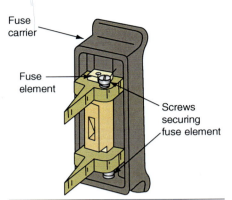

Figure 4.17 Semi-enclosed fuses (BS3036) are sometimes referred to as re-wireable fuses and are cheaper than cartridge fuses and circuit breakers

Did you know? ❓

A major advantage of semi-enclosed fuses is that they are cheap to buy, and it is obvious when the fuse wire has ruptured. However, it takes a lot longer for them to operate in comparison to other protective devices.

Figure 4.19 The main service fuse incorporates silica sand so that it can withstand small surges in the supply

Key safety point

Modern three-pin plugs are designed as a sealed unit in order to reduce the number of people being electrocuted in domestic households.

It is very important to note that of the three overcurrent conditions mentioned previously, an RCD will only operate during condition 3. In other words, an RCD will only operate during earth faults. This is why it is classified as Additional Protection.

Certain circuits and systems however, must have RCD protection in place:

- Such as when cables are not enclosed in metal enclosures such as conduit or similar
- Such as when cables are not enclosed in walls to a depth of 50 mm
- Such as when a circuit's protective device cannot isolate the circuit in time
- Such as when a certain earthing system records high impedance (fault path) values
- For any circuit supplying equipment in bathrooms
- For any circuit supplying socket outlets

Different types of fuses have been fitted within a domestic consumer unit such as a semi-enclosed fuse also referred to as a rewireable fuse. As shown in Figure 4.17, it is effectively an old fashioned type of protective device that is still in use today. Recognised through their British Standard number, BS 3036, they do have advantages, such as being cheap to buy and, once blown, it is obvious that the fuse wire has ruptured. Its main disadvantages are, however, that it takes longer to operate than other types of fuses, or circuit breakers – twice its actual rating, in fact – and because of this a certain correction factor has to be applied when designing circuits.

Cartridge type fuses (Figure 4.18) are also used and differ in design dependant on their use. The main service fuse (BS 88-2) which is located in a domestic cut out (Figure 4.19) is known as a High Rupturing Capacity because it contains silica sand, and it is specifically designed to withstand small surges within the supply. Cartridge fuses are also fitted within consumer units (BS 88-3) as well as plug top fuses (BS 1362). A typical example is shown in Fig 4.18. The ratings of cartridge fuses are labelled on the fuse body which means it should be clear what value is required when replacing a blown fuse. They also tend not to deteriorate over time and the circuit will remain isolated until such time as they are replaced.

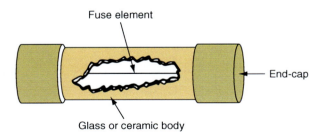

Figure 4.18

The association between certain devices is sometimes shown through their British Standard number. For instance, plug top fuses are numbered (BS 1362) are closely matched to the three pin mains plug top itself (BS 1363).

Although there are in general three different types, domestic consumer units normally use a Type B circuit breaker (Figure 4.20). There are Type B circuit breakers shown in Figure 4.21 – can you spot them? Circuit breakers are more expensive than fuses but their main advantage is that they can react differently to an overload and a short circuit. During an overload situation a circuit breaker

Figure 4.20 Domestic properties normally use Type B circuit breakers

uses a bi-metallic switch which reacts to the excess heat produced and isolates the circuit in a similar way to an electric kettle. Remember, an overload is not a fault situation, but occurs when too much is asked of a circuit, therefore it might take several hours before the circuit breaker actually trips.

A circuit breaker also includes a magnetic trip, so that when excessively large currents flow during fault conditions, the circuit is isolated very quickly. Type C circuit breakers are typically used in commercial or industrial fluorescent lights and incorporate a slight delay in their tripping action. This is needed because fluorescent lights draw a lot of current when starting or striking up. Certain industrial equipment such as X-ray machines, welding equipment and motors also draw excessive currents when first switched on, far more than a Type C can handle. Therefore, Type D circuit breakers were developed to withstand these so-called high in-rush currents.

 Important point

A major advantage of circuit breakers is that they have two ways of operating (thermal and magnetic trip) and can easily be reset. However, resetting a circuit breaker without warning others has also led to electrocution on site.

 Description

Most domestic consumer units house Type B circuit breakers.

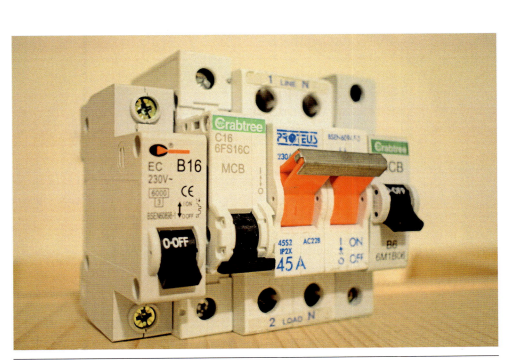

Figure 4.21

Protective device activity

This activity contains two sets of matching terms and each matching pair should be given the same number. The first one has been done for you: 'Fuse = 1' matches with 'Protective device that operates through heat (thermal operation) = 1'

Protective device that operates through heat (thermal operation) = 1	Protective device that operates with either heat or magnetism = 2		Circuit breaker =	Fuse = 1
Additional Protection Device that only operates through magnetism = 3	Type of circuit breaker used in domestic environments = 4		RCD =	Type B =
Type of circuit breaker used with industrial equipment that generate high in-rush currents = 5	Type of circuit breaker used in commercial fluorescent lighting =		Type D =	Type C = 6

As already stated, all domestic circuits should be protected by their own protective device, including being labelled as such within the consumer unit. This assists the user to distinguish which circuit has operated.

Circuits are typically arranged as follows:
- Lighting circuits (installed downstairs and upstairs, cable size: 1.5 mm², protected by a 6 A protective device)
- Ring circuits (installed upstairs, downstairs and kitchen area, cable size: 2.5 mm², protected by a 32 A protective device)
- Dedicated circuits to cater for electric: cookers, showers and water heaters >15 litres

Domestic wiring

Domestic properties are largely installed using a PVC sheathed and insulated cable (Figure 4.22), which use thermoplastic material around the inner cable conductors and its outer sheath. It is very important to realise that the actual sheath is what protects the cable from any potential physical damage and this is known as providing mechanical protection.

PVC sheathed and insulated cable is also commonly called twin and earth as well as PVC flat profile, and it is cheap to produce and is mainly installed in existing walls, floors or through wooden roof joists. Its operating temperature is generally set as 0–70°C.

PVC flat profile cable has two main differences from other cable types and both involve the earthing conductor but its proper title is the circuit protective conductor (c.p.c). This is a very important safety element because any metal

Description

Domestic properties tend to use twin and earth cable, which uses thermoplastic insulation.

part that becomes live through an electrical fault is potentially very dangerous if anybody came into contact with it. Electricity will always take the least path of resistance for itself and the c.p.c offers the fault current an alternative low resistance path to flow down. When this happens, the fault current will increase rapidly, reaching thousands of amps in certain circumstances, which will cause the circuit protective device to operate extremely quickly in order to isolate the circuit and remove any possibility of electrocution. Isolation of the circuit will also stop the cable insulation from melting due to the excessive temperatures that occur during fault conditions.

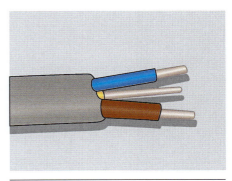

Figure 4.22 Twin & earth, flat profile and PVC/PVC are different names for the same type of cable

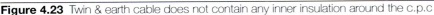

Figure 4.23 Twin & earth cable does not contain any inner insulation around the c.p.c

Because the c.p.c only conducts electricity during fault conditions, all PVC flat profile cables sizes (apart from 1 mm²) are produced with a c.p.c that is smaller in size than the line or neutral conductors. Neither is the c.p.c given any inner insulation; therefore an electrician must fit some earthing sleeving (Figure 4.23).

Duty of care

Electricians do have a duty of care to their customers, but this duty of care should also extend to the quality of the cable involved. Let us examine this through a recent event in Australia.

> **? Did you know?**
>
> Apart from 1 mm², all other twin and earth cables are manufactured with a c.p.c that is smaller in size when compared to the line and neutral conductors.

Case study

This situation, which can be found at the following link -www.basec.org.uk/News/ Basec-News/Multi-million-Dollar-Cable-Recall-in-Australia-Could-Happen-Here - occurred when cables manufactured overseas, which did not meet expected safety standards, were installed in over 40,000 domestic properties in Australia. The insulation used in both the sheathing and insulation became brittle in a very short time period and potentially could initiate an electrical fire.

It is highly likely that this cable, which was sold through major hardware stores, would have been stamped with a certain manufacture code, which indicates it meets their specific manufacturing standards. However, there are other certain markings on cables that demonstrate that it has been independently tested and approved. This applies to all BASEC (British Approvals Service for Cables) because such cables are regularly tested for ageing and many other properties. This situation has led to the installation of 2,500 miles of potentially hazardous cable across five states which is potentially a fire or electrocution threat. There is no information given in relation to the cost of this defective cable, because inexpensive materials are usually an indication that its quality is degraded. This on-going story, however, can stand as an example to other countries, that equipment that is tested independently of the manufacturer and comply with organisations such as BASEC, has passed very stringent quality tests.

General purpose cables

Other cable types used within domestic properties include general purpose PVC white flexible cable, often referred to as flex. This multi-core PVC cable is not as rigid as PVC flat profile and is ideally suited when a piece of equipment requires greater flexibility in its movement, position or operation, such as cabinet lighting displays, washing machines, table and standard lamps and lighting pendants. Because of the possibility of heat damage when cables are routed near sources of heat such as water heaters, thermosetting cables which have an operating range of up to 90°C are normally selected.

Installation sequence

At the beginning of this chapter, the term first fix was mentioned, which applies to the instalment of lighting and power accessories boxes. This is followed by second fix, which refers to the routing of any circuit cables, which are often secured by plastic clips such as that shown in Figure 4.24.

Appendix D of the One-Site Guide contains the recommended spacing distances between supports for cables, conduit and trunking. For PVC sheathed and insulated cables, it requires the installer to obtain the major axis or diameter of the cable, in other words the thickest part. Once this measurement has been established, the horizontal and vertical clipping distances can be obtained by referring to Table 4.1.

This appendix also contains details of the maximum bending radius of cables, so that cables are bent to a suitable radius within acceptable limits (Figure 4.25 (a)) and are not bent beyond a certain amount because this will introduce needless

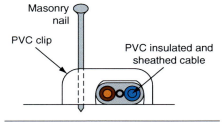

Figure 4.24 Twin & earth cable is secured in place with PVC clips

Table 4.1 Distances required between cable supports for accessible systems

Diameter of cable	PVC thermosetting and thermoplastic		Armoured		Pyro (mineral insulated)	
	Installed horizontally	Installed vertically	Installed horizontally	Installed vertically	Installed horizontally	Installed vertically
Up to 9 mm	250	400	–	–	600	800
Over 9 to 15 mm	300	400	350	450	900	1200
Over 15 to 20 mm	350	450	400	550	1500	2000
Over 20 to 40 mm	400	550	450	600	–	–

Note: When using flat profile cable (twin and earth) use the thickest part of the cable.

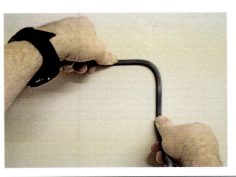 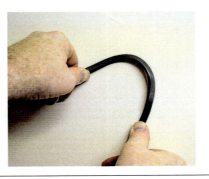 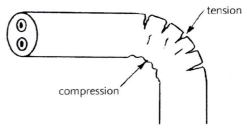

Figure 4.25 (a), (b), (c)

stress by introducing excess tensile and compression forces as shown in Figures 4.25 (b) and (c). Again after obtaining the major axis measurement this is multiplied by a certain factor.

Cables can, in general, be fixed to most surfaces such as brick, wood or plaster but are normally hidden from view and are installed in certain areas or zones as seen in Figure 4.26. This reduces the possibility of a person coming into contact with a live supply.

In general, cables should be laid vertically (up), horizontally (across) but never diagonally. This came about through a tragic case of a person being electrocuted when driving a nail into a wall to hang up a picture. She attempted to secure the picture well away from a socket outlet nearby, assuming that the cables feeding the socket would run up. Unfortunately, the cables had been installed diagonally and tragically she came into contact with a live supply.

> **Key safety point**
>
> Cables should never be installed diagonally.

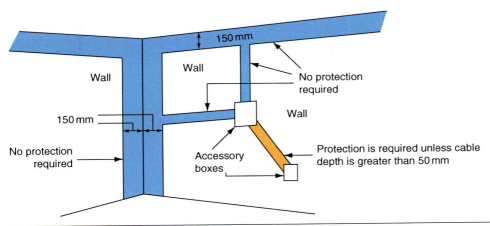

Figure 4.26 In a domestic property cables tend to be installed in certain zones

Lighting circuits are normally wired through what is known as a loop in system, whereby the majority of connections are centralised within a ceiling rose, as shown and disclosed in Figure 4.27.

A supply is fed from the consumer unit and into the ceiling rose in the nearest room. It is then fed out to the next room and so on until all the other rooms within that floor are given a loop in supply. The ceiling rose in each room will then power up each switch by connecting the brown line conductor into the common termination, which is returned back to the ceiling rose through a blue sleeved conductor. However, please note that this particular neutral is no longer

> **Description**
>
> Most of the cables in a loop in system of wiring are terminated at a ceiling rose.

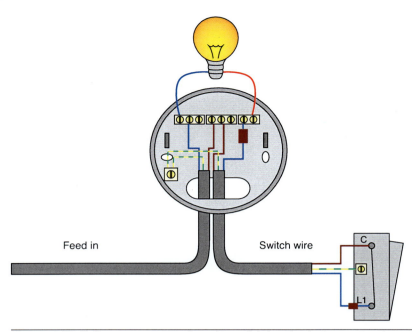

Figure 4.27 Ceiling roses are used to wire up lighting circuits in a system known as loop in

a neutral conductor but is now known as a switch line and should be identified with brown sleeving at both its terminated ends.

Lamps are electrically connected through a lamp holder or pendant either side of the ceiling rose. A one-way lighting circuit or single switch incorporates two connections: Common and L1.

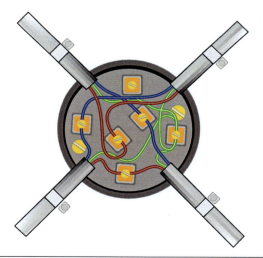

Figure 4.28 Apart from a ceiling rose, lighting circuits can also be wired up with a junction box

Many lighting faults are caused by unqualified or unskilled persons linking all the neutral conductors together, including switched line conductors within a ceiling rose, who are then left wondering why the circuit breaker trips when power is applied.

Another method is sometimes used instead of the loop-in wiring system and connections housed within a junction box (Figure 4.28). The wiring used is the same, in that there is a feed in and feed out, a feed to the switch and a final connection to the lamp holder or pendant.

When additional control is required such as in a stairwell, for example, or when there are multiple doors to a large living room, then two-way lighting is used. This type of switch has three connections: Common, L1 and L2. Figure 4.29 shows this kind of arrangement.

Two-way switch

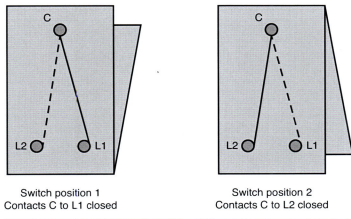

Switch position 1
Contacts C to L1 closed

Switch position 2
Contacts C to L2 closed

Figure 4.29 Two way switching is very common in a stairwell, so that the light can be controlled at the top or at the bottom of the stairs

When wiring a two-way lighting circuit, the basics are very similar to a one-way circuit, in that one of the switches is powered up but thereafter, a three core and c.p.c cable, commonly known as a strapper, is then used to link both switches together as shown in Figure 4.30. This allows either switch to make or break the circuit and each is able to control the light.

Description

Two-way switching utilises a three core and c.p.c but is also referred to as a strapper.

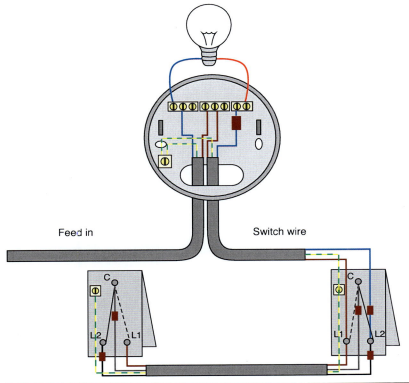

Feed in

Switch wire

Figure 4.30 A strapper cable straps the supply between two switches

An intermediate switch on the other hand, which is shown in Figure 4.31, has four connections and is used to control lighting in long corridors or flats that contain multiple floors. However, an intermediate switch has to link all six of the strapper connections, therefore it requires the use of a strip connector block to connect one of the pairs of conductors.

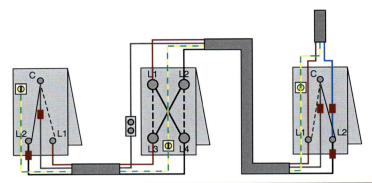

Figure 4.31 An intermediate switch has 4 connections (2 in, 2 out)

Key safety point

Bathrooms are considered as a special location, and lighting is normally controlled through a pull switch.

Finally, because of the possibility that a person visiting a bathroom or toilet may have wet hands, lighting in this kind of environment is normally controlled through a pull cord.

It is worth reinforcing that every electrical connection, irrespective of how it is made, must be both electrically and mechanically sound. Electrically sound means that the termination is very low in resistance. This is because badly formed terminations cause a high resistance joint and are the main cause of electrical fires.

Lighting

There are many different types of light fittings used in domestic properties, ranging from General Light Service – a basic light bulb (Figure 4.32), fluorescent lamps also known as low pressure mercury (Figure 4.33), compact fluorescent lamps (energy saving, Figure 4.34), light emitting diodes (LEDs, Figure 4.35) and tungsten halogen (Figure 4.36).

Figure 4.32 General Light Service (GLS)

Figure 4.33 Fluorescent lamp (low pressure mercury)

Figure 4.34 Compact fluorescent lamp (energy efficient)

All these lights operate differently – for instance, the GLS produces light through an internal heater which is why it is classified as an incandescent lamp. This simply means 'white hot'. They are generally very cheap to buy but are not very efficient or long lasting. Because of their relatively inefficient nature, GLS have

Figure 4.35 Light-emitting diode (LED)

Figure 4.36 Tungsten halogen lamp

been gradually phased out and are being replaced by far more efficient lamps such as compact fluorescent lamps.

Flourescent type lamps produce light by passing electricity through a gas (low pressure mercury). They do cost more than GLS lamps, but are more efficient and last a lot longer. Fluorescent lamps, however, require extra components to work, such as a starter to create a spark and a choke to generate a very large voltage spike. A compact fluorescent lamp is a type of mini fluorescent spotlight but some people complain that they take an age to come up to full brightness.

Tungsten halogen is another variety of incandescent lamp, which uses a tungsten gas to create a chemical reaction that actually redeposits bits of tungsten, which helps give the lamp longer life. LEDs on the other hand operate through a principle known as electroluminance, which is when a material emits light after electricity is passed through it. LEDs do cost quite a bit more to purchase, but last a lot longer than other lamps and are also far more efficient in terms of how much electricity they draw when operating.

Tackle a difficult word: incandescent

White hot

? Did you know?

Certain ratings of GLS lamps were discontinued because they were very inefficient and bad for the environment.

? Did you know?

The choke used to produce a large voltage strike in a fluorescent lamp works on the same principle as a coil which is fitted within a car's ignition system.

Figure 4.37 Recessed lights are sometimes referred to as downlighters

A lot of houses these days opt for recessed lighting known as downlighters (Figure 4.37) which can be fitted with either tungsten halogen or LED lamps, as long as they are manufactured as GU 9/10 assemblies.

Socket outlets

All socket outlets in a domestic property should incorporate a switch (switched) and are normally supplied as either single or double, although others are

Figure 4.38

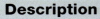**Description**

Ring and radial are types of electrical power circuits.

available. Newly designed socket outlets can these days contain USB ports, which makes recharging mobile devices that much easier. The number of sockets in an outlet is sometimes indicated through the term 'gang', as shown by the two gang socket outlet in Figure 4.38.

There are generally two ways of wiring up power circuits and they are called ring and radial. A radial circuit feeds each socket outlet in a particular circuit but stops at the last one connected in that run. A ring on the other hand, once it has fed the last socket, is returned to the consumer unit and back to the same protective device as specified in Figure 4.39.

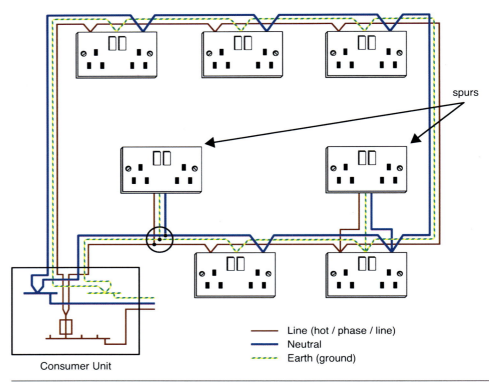

spurs

Line (hot / phase / line)
Neutral
Earth (ground)

Consumer Unit

Figure 4.39 Each socket in a ring circuit can spur one additional unfused single or double socket

Domestic properties tend to use three ring circuits, so that power is distributed to:

- Downstairs sockets
- Upstairs sockets
- Kitchen area

A kitchen is provided with its own ring circuit because it normally contains certain equipment which requires a lot of electricity to operate. Kitchens typically include:

- Washing machine
- Tumble dryer
- Fridge freezer
- Electric fryers
- Coffee makers
- Electric blenders

Figure 4.40 Kitchens tend to house a lot of electrical equipment, which is why it's given its own ring circuit

A kitchen will also include items such as kettles and toasters, which are designed to draw a lot of current in a short space of time. If all these items were added to the downstairs ring circuit then they could overload it.

For the same reason, because the amount of current drawn by cookers, showers or large water heaters is potentially very large, they are all supplied through their own dedicated radial circuit.

Modern mains operated equipment is powered through a BS 1363 plug (Figure 4.42), with most appliances tending to incorporate sealed plugs. This change

? Did you know?

At one time 20 deaths were recorded every year in the UK that resulted simply from incorrectly wiring a three-pin plug.

Figure 4.41 Always act fast when you see signs of over heating

Figure 4.42 The cable clamp, should always be tightened down on the sheath

was introduced to reduce the number of non-skilled persons attempting to work on three-pin plugs.

Safety is also a prominent feature in socket outlets that are installed in bathrooms in order to power electric shavers. Using any electric device when your body is possibly very wet is potentially hazardous, because water reduces a person's body resistance, which increases the effect of an electric shock. This is why a bathroom is known as a Special Location and consequently a shaver socket is powered through a safety isolation transformer.

The On-Site Guide contains an appendix, which shows how certain domestic circuits are standardised in size, including the rating of their protective device. These details are shown in Table 4.2 and it is important to realise that although any number of sockets can be fitted within a ring circuit, it cannot supply a room any bigger than 100 m² in area.

Table 4.2 General requirements for standard circuits

	Type of circuit	Overcurrent protection device	Minimum conductor size (mm²)	Maximum floor area (m²)
A1	Ring 30 or 32	Any type of device	2.5	100
A2	Radial 30 or 32	Cartridge fuse or circuit breaker	4	75
A3	Radial 20	Any type of device	2.5	50

Basic protection

Insulation due to its properties alongside the use of enclosures (encasing live parts), obstacles (form a barrier around live parts) and placing supplies out of reach are all examples of how a person is protected from coming into contact with live supplies. Such measures are known as providing 'basic protection'.

Fault protection

If basic protection stops accidental contact with electricity, fault protection tries to limit any damage that a fault condition can cause by isolating the circuit. Any metal component within an electrical system which a person could come into contact with, such as that shown in Figure 4.43, is known as an exposed conductive part and must be 'earthed' through a c.p.c. All the protective conductors (c.p.c) are in turn connected to the earthing bar within the consumer unit. The main earthing conductor which should be 16 mm² in size, completes the link between the earthing terminal and the incoming supply. When a fault current occurs, it actually flows back to the supply transformer and, in general, there are three common earthing systems that make this happen.

- TNS (earthing is achieved by using the main supply armouring, rather than a separate conductor)
- TNCS (c.p.c and neutral conductor are linked and combined at the transformer as well within the supply cut out, but are separated throughout the property)
- TT (both the transformer and the property are earthed directly to the ground by using an earth electrode)

Description

Earthing is associated with all exposed conductive parts that form part of an electrical installation, for example, a metal plate switch.

When an earth fault occurs it follows the following sequence:

- Exposed conductive part (metal switch box) becomes live
- Fault current flows down the c.p.c into the earthing terminal
- Fault current flows through the main earthing conductor and back to the supply transformer either through the supply armouring, neutral conductor or through the earth itself

Figure 4.43 All parts of an installation that are conductive (metallic) are known as exposed conductive parts and have to be earthed

Earthing is so important that I have again summarised its main aims:

- Earthing provides a low resistance path
- Electricity always flows down the path of least resistance
- This means if given a choice it will flow to earth rather than through a person
- During a fault situation, the actual current rises rapidly
- Protective devices such as a fuse or circuit breaker will operate and isolate the circuit
- If fitted, an RCD will operate even quicker than other protective devices, but only if the fault current flows down to earth

All equipment that requires earthing is classified as Class 1, but there is also a certain type of equipment which is known as Class 2 or double insulation. In this type of equipment, any live parts are encased within two barriers or layers of insulation, therefore it should not be physically possible for any person to come into contact with any live supply. Because of this physical barrier, Class 2 equipment does not require an earth connection. A shaver socket installed in a bathroom is a perfect example; given its location, isolating live parts from physical touch is a better means of protection than earthing. The shaver socket is also powered through a safety isolation transformer; the end user (person shaving) is isolated from the incoming supply because internally a transformer operates through magnetism.

Key safety point

Earthing provides a low resistance path for a fault current to flow down. The fault current will rapidly increase in value and a protective device will operate to isolate the circuit.

Description

Bonding is associated with all metal extraneous conductive parts that do **not** form part of an electrical installation, for example, a metal gas pipe.

There are, however, other items, and especially other services such as gas and water, which are supplied from metal pipes. The big difference here is that although they are not there to provide electricity they are often made from metal – we still have to provide protection just in case a live conductor accidentally comes in contact with them. Anything that is made from metal but does not form part of an electrical circuit is known as an extraneous conductive part. I think of it as what **extra**(neous) bits of metal do – I have to protect them from becoming live. This protection is known as equipotential bonding and it too provides fault protection.

Bonding is carried out by connecting a 10 mm² main equipotential bonding conductor from the earthing terminal within the consumer unit and terminating it around water and gas pipes through an earth clamp (Figure 4.44) which needs to be compliant with British Standard: BS951. A typical earthing and bonding arrangement is shown in Figure 4.45.

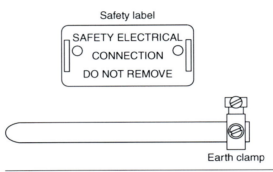

Figure 4.44

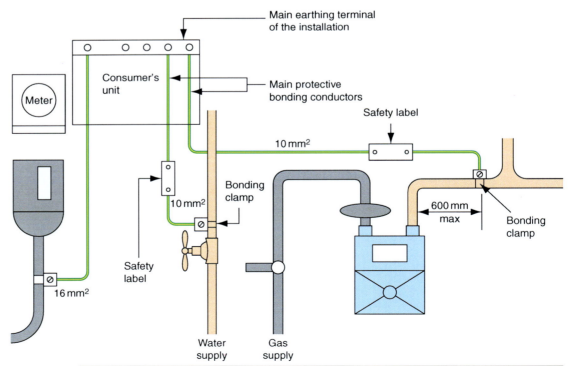

Figure 4.45 Gas and water pipes are known as extraneous conductive parts and have to be bonded

Time out, let's summarise some **key** points!

- In most domestic properties, a 230 V single phase supply is sufficient to generate enough current to meet the needs of the electrical equipment
- The main supply is normally delivered through a PVC steel wire armoured cable, designed specially for use underground and is very strong due to its metal armouring
- A domestic supply is normally terminated at the service head
- The wiring used tends to be PVC flat profile cable because it is cheap to manufacture and is usually embedded in the walls or routed through wooden floorboards or joists
- Three ring circuits are normally installed upstairs, downstairs and in the kitchen area
- Wiring is also installed in set zones, being placed vertically and horizontally but never diagonally
- RCDs installed give additional protection against electric shock, but cannot be the sole means of protection
- Fuses and circuit breakers will operate during fault conditions because an earthing system ensures that it provides a low resistance path
- The earthing system allows a very large current to flow and operate any protective device very quickly

To review your understanding try this True or False quiz based on electrical installation terms and knowledge.

Question 1: PVC flat profile cables are secured in place using PVC clips.

True	False

Question 2: Recessed lights are known as downlighters and are normally fitted with either tungsten halogen or LED lamps.

True	False

Question 3: A metal gas pipe is known as an exposed conductive part.

True	False

Question 4: Earthing provides a low resistance path for a fault current to flow, which will cause a protective device to operate very quickly and isolate the circuit.

True	False

Figure 4.46

A journey around a commercial property

Commercial properties include shops, restaurants, offices and even classrooms, and they are generally a lot different from a domestic property, although there are some similarities. For instance the incoming supply will be again be supplied through a PVC/SWA cable and it will be terminated in a distribution board. However, because the amount of current required is far greater than that supplied to a domestic property, then it requires a three-phase supply which is this time terminated at a three-phase distribution board.

The delivery system will also be different since it is common to install commercial outlets in plastic trunking (square or rectangular) and/or conduit tubing (Figure 4.46). Both are referred to as wiring enclosures.

Because the actual enclosure itself provides mechanical protection, PVC unsheathed wiring is used although it is more commonly referred to as 'singles'.

This means that every line, neutral or earth conductor, will comprise of a single cable but it's very important to realise that they should never be used outside of metal or plastic enclosures. Singles should be drawn off from cable reels as shown in Figure 4.47 and installed in fairly short sections rather than at the end of a long installation run. It is also good practice to draw all the conductors through together to avoid any possible damage occurring.

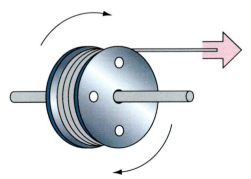

Cable *run off* will not twist; a short length of conduit can be used as an axle for the cable drum

Cables allowed to *spiral off* a drum will become twisted

Figure 4.47

Power is normally delivered through socket outlets that are housed within the trunking, or positioned nearby and fed directly through conduit. Mini-trunking (Figure 4.48) can also be used to cover small runs of cable either within commercial or even domestic properties.

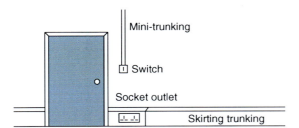

Figure 4.48

There is also an established conduit method of wiring up lighting systems as indicated below in Figure 4.49.

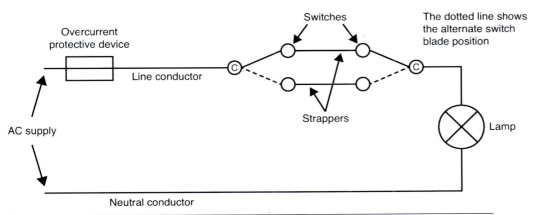

Figure 4.49 Wiring enclosures such as conduit, are wired with unsheathed cables referred to as singles

In certain commercial locations, low smoke and fume cables are used in lighting supplies, especially in public areas. This is because LSF cables emit far less smoke and corrosive gases in comparison to thermoplastic insulation.

Other locations such as a commercial kitchen or plant room, demand greater protection from impact damage or heat and therefore metal wiring enclosures are far more appropriate. If corrosion is an issue or possibility, then galvanised metal conduit which incorporates an additional zinc coating would be preferred.

A journey around a farm (agricultural)

Farms are very dangerous environments because they contain lots of machinery, large, powerful animals, chemicals, dirt and grime and even attract a lot of rats. Therefore the wiring system needs to be appropriate for such an environment. If the farm involves milking parlours, for instance, then it will require a lot more current than a single phase circuit can supply, therefore again this kind of installation requires a three phase and neutral supply. Because of the high possibility of mechanical damage from the machinery, and even the animals, then the delivery system, especially when installed outdoors, is typically through

Important point

Animals are very vulnerable to the effects of an electric current.

galvanised steel conduit. However, if such metal wiring enclosures were ever to become live they would pose a serious threat to people, but especially to animals. This is because most farm animals such as horses, cows and sheep are very vulnerable to electric shock and can die if they come into contact with as little as 25 V.

Let's examine this further and find out why.

- A horse sometimes wears metal shoes
- Animals sometimes stand directly on wet ground
- Large animals have very low body resistance
- Because they stand on four feet it creates a big potential difference between their front and back legs, which allows current to travel straight through the animals heart

This is why heavy duty plastic conduit is sometimes used on farms, since it is non-metallic, and far stronger than ordinary conduit. Given the environment, it is also easier to clean.

Figure 4.50 Animals are very susceptible to even very low values of electric shock

Electric fences (Figure 4.50) are, however, used to control animals, but the controllers used tend to pulse any electric shock so that their effect is limited and not damaging.

There are other issues with installing plastic enclosures; for instance, it must be warmed up to room temperature before it can be bent otherwise it can crack. Heat and solar radiation from the sun is also a consideration when it's installed outside, because this can cause plastic to expand. To counter this, expansion couplers should be fitted every five metres to take up this expansion.

A journey around a factory

Most industrial locations, such as factories, will include machinery that require a lot of electricity to operate, far more than most commercial properties. This means that again a three-phase supply is required, supplied through a PVC/SWA sub-main and terminated in a three-phase distribution board. Dependant on the amount of current required, industrial factories can make use of Power Track (Figure 4.51) or a bus-bar system, which uses several metallic strips rather than actual cables. Bus-bar systems are also useful because they allow different sections of the factory to be controlled and isolated through what is referred to as switch fuse isolators. This means that when essential maintenance is required within one section of a factory other areas are not affected.

Industrial environments also use both metal conduit and trunking extensively given the high degree of mechanical protection that they provide.

When rigid metal conduit is used it needs to be supported by various fittings shown in Figure 4.52 and range from distance saddles which are used on uneven surfaces and crampets which are usually plastered over. Alternatively, in other locations due to the hazards involved from machinery and assembly lines, the electrical supply is routed away from the factory floor and supported from overhead chains.

SWA cables are also used extensively in industry, but are very heavy, especially in large sizes. Because of this, certain containment systems such as steel tray (Figure 4.53) and ladder (Figure 4.54) are fabricated to offer much needed support.

Important point

A SWA cable offers a lot of mechanical protection due to its metal armouring.

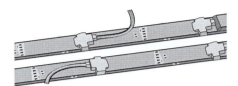

Figure 4.51

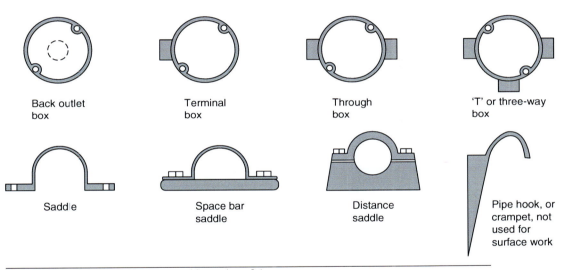

Back outlet box

Terminal box

Through box

'T' or three-way box

Saddle

Space bar saddle

Distance saddle

Pipe hook, or crampet, not used for surface work

Figure 4.52 Conduit is also secured by various fixings

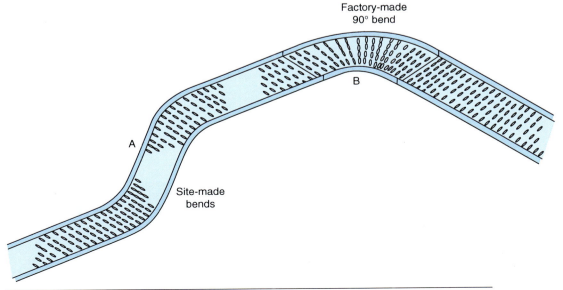

Factory-made 90° bend

B

A

Site-made bends

Figure 4.53 Tray is used to support SWA cables

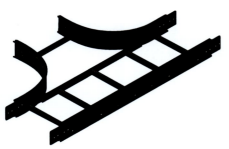

Figure 4.54 Apart from tray, ladder is also used to support SWA cables

Cables are also used to provide Safety Services such as providing alarms systems and emergency lights. The cables in use are not as heavy as SWA, therefore cable systems are routed and supported on a containment system called basket (Figure 4.55) which does not have to be as rigid as metal tray.

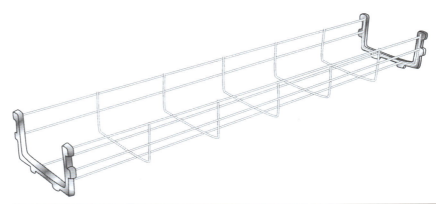

Figure 4.55 Basket is a containment system designed to support data and emergency cables

Electrical systems can operate on different voltages; for instance, Band I is known as extra low voltage and covers values up to 50 V. Band I voltages are deemed too low to provide a serious electric shock and include telecommunications and control systems.

Band II ranges from 50 V to 1000 V and largely covers the voltages used in electrical installation.

In certain circumstances if cables from both Band I and II are installed together, higher voltage cables could cause interference through certain electromagnetic effects that are induced on to lower voltage band systems. Designers are faced with two choices, either they use multi-compartment metal trunking as shown in Figure 4.56, which not only separates Band I and II cables but is also used to allocate a separate compartment for any cabling associated with emergency supplies. Alternatively, single compartment trunking can be used, but all cables within it must be insulated to the highest voltage present.

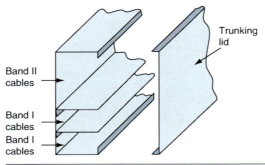

Trunking lid

Band II cables

Band I cables

Band I cables

Figure 4.56 Multi-compartment metallic trunking is excellent to stop electromagnetic interference that could otherwise occur from different voltage systems

Important point

Flexible conduit can withstand any excessive vibration produced by electrical motors.

Plastic and metal conduit are both fairly rigid in design and not very bendable, which means they are not very appropriate to supply electrical motors which physically vibrate when operating. Flexible conduit shown in Figure 4.57 was introduced to provide the final connection to electrical machines such as motors, since it can withstand and absorb any vibration safely.

Wiring systems activity

This activity contains two sets of matching terms and each matching pair should be given the same number. Enter in a number that matches the correct term. The first one has been done for you: 'PVC Flat profile = 1' matches up with 'Domestic wiring = 1'

PVC flat profile = 1	PVC/SWA = 2	Domestic wiring = 1	Sub-main =
Flexible conduit =	Tray/ladder =	Used with motors to offset vibration = 3	Metal systems used to support SWA cables = 4
Plastic conduit = 5	Metal conduit = 6	Type of enclosure used in commercial environments =	Type of enclosure used in industrial environments =

Figure 4.57 Electric motors often vibrate whilst operating, which is why their final connections are often wired with flexible conduit

Time out, let's summarise some **key** points!

- Different types of electrical installations are used in domestic, commercial and industrial environments
- Different external effects such as impact damage, water ingress, high temperatures, possible attack from animals or plants will dictate what kind of electrical installation system will be installed
- In most domestic properties, a 230 V single phase supply is sufficient to generate enough current to meet the needs of the electrical equipment
- Domestic properties tend to be wired with PVC flat profile cable because it is cheap to manufacture and is usually embedded in the walls or routed through wooden floorboards or joists
- Commercial environments, typically offices, classrooms but also restaurants, use plastic wiring enclosure systems such as plastic trunking and conduit
- Because of the amount of electrical current required by commercial and industrial environments, they are usually fed by three-phase supplies
- In some industrial environments, electrical supplies can be fed from bus-bar systems, metal wiring enclosures or alternatively containment systems such as ladder or tray

Try this True or False quiz to review your understanding based on industrial and commercial environments.

Question 1: Plastic conduit is far stronger than metal conduit.

True	False

Question 2: Industrial environments often route electrical cables overhead and away from the floor area.

True	False

Question 3: Conduit is wired through non-sheathed PVC singles.

True	False

Learning outcome 2.4

State the items to look for when checking finished work.

After an installation has been completed it must be both inspected and tested to ensure that it meets the requirements of the designer and also the Wiring Regulations (BS 7671).

Inspection stage

Prior to testing, an inspection stage must be carried out and done correctly, it makes use of many of our human senses such as: sight, hearing, smell and touch. The inspection stage will look closely to ensure that accessories and enclosures are not in any way damaged and that cables are protected, laid and routed correctly. Visual checks should also ensure that grommets have been fitted in and around enclosure entry points, otherwise damage to the sheath and inner insulation of cables is possible. Inspection should also, when possible, ensure that terminations do not show excessive amounts of copper and that they are mechanically sound, including being able to withstand someone tugging at the connections.

Dead tests

Dead tests are performed to ensure that there are no potential dangers in place before any circuit is made live. It will ensure that cable insulation is healthy and that safety measures such as earthing is in place. Although the tests involved can be carried out by a multi-meter (Figure 4.58), it is important to remember that there are three distinct elements which must be carried out as follows:

* Continuity of protective conductors
* Insulation resistance
* Polarity

Let us know examine each of these in turn.

Continuity of protective conductors

To fully understand this aspect we must review what is again meant by fault protection and especially the term earthing. All metal parts of an installation that can be accessible by touch are known as exposed conductive parts. These

Important point

When inspecting electrical circuits many of the human senses are used such as using touch to feel the outer sheath of a cable to ensure it is not damaged.

Figure 4.58 A multi-function meter is often referred to as a multi-meter

metal parts are therefore very dangerous if, through a fault condition, they were to become live and a person came into contact with them. As explained earlier, exposed conductive parts have to be earthed through a protective conductor known as the c.p.c, which offers a low resistance path to any fault current that develops. Continuity of protective conductors is a test to ensure that all circuit protective conductors are connected together and continuously throughout the circuit. When measured it should record a very low value of resistance, typically 0.05–0.08 Ω.

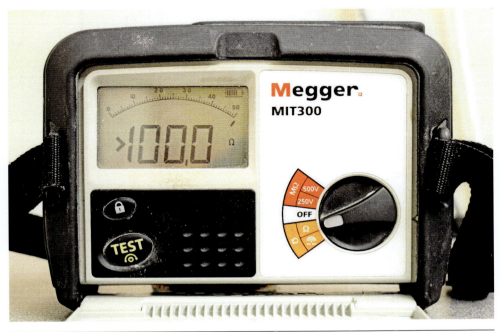

Figure 4.59 The Ω sign indicates Resistance

Procedure

- Set the instrument to the ohms scale Ω (Figure 4.59)

When measuring the resistance of a length of cable, we must take into account the resistance of the leads by subtracting it from the total reading of the conductor under test.

- Results should always be recorded in ohms and, as stated previously, healthy values should range between 0.08–0.05 Ω.

Insulation resistance (IR)

An IR test is carried out to ensure that the actual insulation material that surrounds the conductors is healthy and affective. In other words the insulation material is **not** breaking down or indicating a short circuit condition, by recording very high values of insulation resistance.

Procedure

- Remove all loads, especially sensitive loads such as dimmer switches
- Set the instrument as indicated in Figure 4.60 to the megohm scale MΩ (millions of ohms)
- Set test voltage to 500 V approximately twice the working voltage, again indicated in Figure 4.60
- Readings should be at least 1 MΩ.

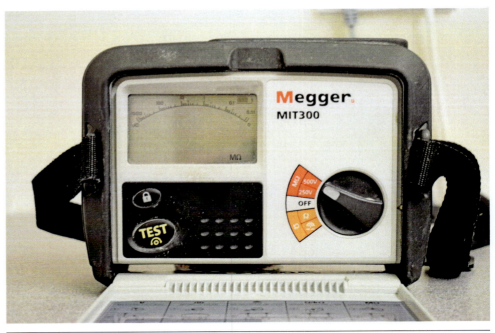

Figure 4.60 The M Ω sign indicates Insulation Resistance

Polarity

Testing the polarity of circuit involves ensuring that all switches and protective devices are fitted in the line conductor only.

Testing activity

This activity contains two sets of matching terms and each matching pair should be given the same number. The first one has been done for you: 'Earthing = 1' matches with 'Circuit Protective Conductor = 1'

Earthing = 1	Exposed conductive part = 2		Circuit protective conductor = 1	Metal switch box =
Healthy value for continuity of c.p.c. =	Minimum acceptable insulation resistance value =		0.05 Ω = 3	1 MΩ = 4

Learning outcome 2.5

Recognise documents which affect installation of wiring systems.

The term non-statutory regulation was raised earlier in Chapter 1 and refers to documents that, although they are not legal in themselves, are written in such a way that all legal aspects are met. For instance, the Wiring Regulations are not themselves legal, but by always adhering to them an electrician is following what

Important point

Insulation resistance should return very high values such as >1 MΩ.

Important point

The Wiring Regulations and Building Regulations are not legal documents themselves, but form what is called a code of practice.

is known as a Code of Practice (CoP). This is why they are sometimes called the Electrician's Bible. They also are given a British Standard number (BS 7671) so that all electricians in Britain follow the same procedures and standards of practice. BS 7671, or as they are commonly known, the 'Regs', are very detailed and therefore a small companion booklet was developed to help an electrician apply the regulations on-site. The booklet is therefore suitably called 'The On-Site Guide' and actually contains extra information not contained in the Wiring Regulations such as clipping distances of cable supports, certain images in relation to dedicated wiring zones as well as the recommended heights of switches and sockets.

A lot of this information is actually drawn from The Building Regulations (2010 as amended) which is a set of rules that applies to new buildings or new extensions to ensure that they comply with certain requirements.

The scope of each Approved Document is given below:

Part A   structure
Part B   fire safety
Part C   site preparation and resistance to moisture
Part D   toxic substances
Part E   resistance to the passage of sound
Part F   ventilation
Part G   hygiene
Part H   drainage and waste disposal
Part J   combustion appliances and fuel storage systems
Part K   protection from falling, collision and impact
Part L   conservation of fuel and power
Part M   access and facilities for disabled people
Part N   glazing and facilities for disabled people opening and cleaning
Part P   electrical safety
Part Q   security - dwellings (England only)

Each aspect is driven by an Approved Document to ensure that important systems such as electrical safety and fire safety are correctly installed in order to protect the occupants. Other Approved Documents consider the impact the building will have on the environment as well as the neighbours by installing sound insulation to reduce the amount of noise produced.

The heights of switches and sockets as shown in Figure 4.61 is actually driven from Approved Document M, which ensures that those designing services inside a building consider the less able amongst us.

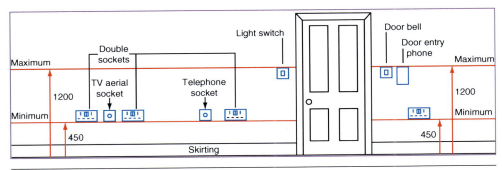

Figure 4.61

Other non-statutory information that is particularly useful to an electrician includes GS38, a HSE publication which acts like a method statement for electricians indicating best practice when carrying out safe isolation procedures. Lastly, the Institution of Engineering Technology, who publish the Wiring Regulations, also publish a range of guidance notes, with GN 3 providing additional information in relation to inspection and testing.

Time out, let's summarise some **key** points!

- After an installation has been completed it must be both inspected and tested to ensure that it is complete and safe to use
- Although it will be tested, the inspection aspect is also very important, and several human senses such as sight, touch and hearing can alert us to potential problems
- An electrician should always comply with their Code of Practice and this is done by always referring to the Wiring Regulations (BS 7671)
- Although BS 7671 is not a legal document, it is written in such a way that it complies with all legal documents
- A low resistance ohmmeter is used to measure continuity of conductors and should register a very low value such as 0.05 Ω
- An IR meter is used to ensure that the insulation resistance of a cable is significantly high by returning values >1 MΩ

To review your understanding, try this True or False quiz based on documentation.

Question 1: BS 7671, the Wiring Regulations are legal documents.

True	False

Question 2: If you are building a new house, the Building Regulations can be ignored.

True	False

Question 3: By always referring to BS 7671 an electrician is complying with their code of practice.

True	False

Learning outcome 3.1

Recognise the main roles of the site team.

Construction project

When a contract is awarded to build large-scale projects, such as a new school or housing developments, the actual process will involve several stages. We will now take a journey around a building site and explore the roles of the various trades working together, who are ultimately responsible for different aspects and services.

The actual contract would have gone through what is known as a tendering process, in that different building contractors would submit closed and sealed bids, which contain an amount or price they have put forward to carry out the build. The building contractor will also work closely with the architect who is normally responsible for producing detailed plans of what the client requires. For a new housing estate the client could be a property developer or a major building contractor, but any public construction such as a new school, in most cases, would also involve the local council and local education authority. Unless qualified to do so, the architect would also normally consult with specialist engineers regarding electrical, water and gas supplies, in order to ensure that the installation complies with all current regulations and requirements.

Once the main contract has been awarded, the chosen building contractor will seek other smaller and more specialised firms to sub-contract specific trade-based aspects. Sub-contractors normally include bricklayers, joiners, plasterers, tilers, roofers, plumbers, gas fitters and last but not least electricians.

A certain amount of groundwork will need to be carried out before anything substantial can be built, and this is typically carried out by ground workers. This phase requires that the area is pegged out and cleared for use so that its foundation can be laid, as well as drainage and other pipework. Plumbers will also be involved at this time in order to outline where drainage and primary services such as gas and water are to be located and situated. Thereafter, bricklayers will be responsible for erecting the actual structure of the building, normally through bricks, blocks or even a combination. Before all this, however, it is common for electricians to set up temporary electrical supplies so that all hand-held equipment are fed through a 110 V reduced voltage system.

Once the shell of the building has been established, carpenters will erect wooden partitions and doorframes alongside the roof to shield the building from the elements. When dealing with large-scale domestic property development, wooden partitions, staircases as well as roof trusses are normally designed and put together off-site by a separate contractor and then delivered to the property. For a school, however, most staircases are either made from wood or even stone in some cases.

Once these partitions are in place, plasterers will fit and apply sheets of plasterboard which are secured by galvanised nails. Plumbers and electricians will both carry out their first fix, electricians positioning various accessory boxes, while plumbers will be engaged in assembling and installing essential pipework.

Plasterers would soon thereafter begin to apply a first coat of plaster called a scratch coat, but they need to work very closely with other contractors since conflict can occur when other tradespeople require access to the same work

Description

A tender is a process by which a company submits a closed bid that represents how much money they require to carry out certain specialist work.

Description

A clerk of works is a person who oversees building work in progress. In other words, they are the eyes and ears of the architect on site.

Important point

Once a new building is water proof, joiners, plumbers and electricians will begin to install their respective systems.

Description

The first coat of plaster is referred to as a scratch coat.

area. For example, electricians will carry out their second fix and lay and secure cables in place, although they are not terminated at this time. Not until this stage is complete can the wall be sealed with an intermediate (middle) layer of plaster.

Once the final layers of plaster have been applied, usually commonly called pink due to its colour, electrical and plumbing final fix will terminate all cables and pipes respectively. Joiners will also finish off other wooden or plastic fittings such as architrave and skirting as well as windows and doors. Joiners can also be involved in fitting out kitchens and utility rooms with purpose built units.

The installation of a new school is called a commercial build, which means that electrical systems will typically be fed through plastic trunking and conduit. That said, in certain locations, such as a boiler house, metal versions will be used because they are far stronger and more robust when considering possible impact damage as well possible high temperatures that are likely to be experienced. Other electrical supplies are fed through what is called bench trunking, such as that found in laboratories.

Once all the internal infrastructure and services such as water, gas and electricity are in place, having of course been inspected and tested by skilled persons, then other trades such as tilers will decorate certain locations such as kitchens, toilet blocks and shower rooms. Any painting is normally carried out by specialist decorators that initially coat walls and ceilings with undercoat and perhaps two coats or more of paint as well as applying different grades of varnish to wooden surfaces.

Because of the sheer number of operations going on within any building site and the multiple trades working alongside each other at any one time, such environments tend to be managed by site managers and supervisors. Site managers are responsible for the day-to-day operations on site and they will coordinate site safety briefings and induction programmes. Site supervisors will share in the responsibility of running a construction site, especially in ensuring that health and safety policies and procedures are adhered to.

Roles in the construction industry activity

This activity contains two sets of matching terms and each matching pair should be given the same number. The first one has been done for you: 'Electrician = 4' matches with 'Craftsperson who installs electrical systems = 4'

Architect = 1	Clerk of works = 2	The eyes and ears of the architect on site =	Person who creates drawings and plans =
Joiner =	Electrician =	Craftsperson who builds things by joining pieces of wood = 3	Craftsperson who installs electrical systems = 4

Plumber = 5	Gas fitter = 6	Craftsperson who installs gas fires, cookers and boilers =	Craftsperson who installs drinking water, sewage and drainage systems =
Plasterer =	Ground worker =	Craftsperson who is employed to prepare a site for the shallow foundation of a new home = 7	Craftsperson who forms a layer of plaster on an interior wall or ceilings = 8
Tiler =	Decorator =	Craftsperson who tiles out kitchens and bathrooms = 10	Craftsperson who applies paint and wall paper = 9

Learning outcome 3.2

State the importance of company policies and procedures that affect working relationships.

It is vital that everyone on a construction site is aware of their role and responsibilities and this is true of both employers and employees. This is because everyone has a duty of care to each other and failing to carry out your job correctly could not only injure yourself but also cause injury to your fellow workers and even possibly any passers-by.

This is why companies carry out induction programmes, so that all workers are aware of company policies and procedures, especially to define:

* What you can expect from your company
* What your company expects from you

The induction programme will also indicate the level of behaviour and attitude expected when working on a construction site, because with so many operations going on at any one time, it is an extremely hazardous environment, so much so that it too is classified as a 'special location'.

Timekeeping is also important; you are expected to arrive on time and be in a fit state to work. Being consistently late is a disciplinary matter and ultimately serial offenders can be, and often are, dismissed. Employers have a duty of care to provide adequate supervision to non-qualified people, including apprentices. There are also other situations that require the positioning of a safety person, posted to monitor a colleague who might be working at height or even measuring live electrical supplies. Being continually absent therefore puts pressure on employers to cover all on-going situations and they will soon become tired of having to provide alternative personnel when the same person is continually sick or absent. These kind of delays can also have a knock-on effect regarding the overall work programme which is scheduled to meet certain targets and dates of completion. This can lead, in certain circumstances, to heavy fines for your company, because penalty clauses are sometimes written into contracts.

 Description

An induction programme ensures everyone on site is aware of their responsibilities as well as certain safety procedures.

When working on a construction site there is an expectation that all workers comply with a standard dress-code. In fact, without a hard hat, high visibility vest and safety boots employees will simply not be granted access. Certain areas will be particularly hazardous and therefore off-limits to most employees. Access will be strictly controlled - only authorised personnel may be permitted entry. In addition, yellow warning and blue mandatory signs (Figure 4.62) will be positioned to highlight the existence of any particular hazard and if any specific protective equipment must be worn whilst working in or migrating through these areas.

Figure 4.62 Site Safety is often promoted through various signs

Employees should always embrace a positive attitude when performing their duties, especially since safety officers, site supervisors and managers do patrol construction sites. If they witnessed an employee who is blatantly not adhering to site procedure and putting themselves or others at risk, then they can be issued a yellow card. This effectively stops them from continuing with their task and disciplinary procedures could follow.

Case study

Every single RAF station I was posted to had such an induction programme, which lasted for up to a week in certain circumstances. The roles of aircraft trades are widespread and can range from working on fighter aircraft, bombers and even passenger airlines such as the VC 10 and although each station might vary a little, they all had concrete and rigorous processes designed to underpin safety.

Compare and contrast this situation to another company that I worked for, just before I enlisted, which was also involved in carrying passengers. In 1980, I worked for a famous train company as a steam engine fireman, employed to stoke the fire with coal. A difficult job at the best of times, since for fear of losing too much heat, the fire door could only be open momentarily and in that time, I had to dig down and shovel the coal in. The coal was always shovelled to the left or right of the fire plate, because the sheer vibration helped spread the coal evenly over the surface. All this was explained by the driver, as well as other duties such as how to clean the fire without extinguishing it and so forth. However, there was no mention of any safety briefs, no coverage of how to move the train but more importantly how to stop it in an emergency. As far as the driver was concerned, if he told me everything he knew, then I would know as much as him and that would never do. In other words, the driver wanted to keep the knowledge in order to keep the power.

Another fireman took it upon himself to demonstrate how to take control of the train in an emergency. I was now armed with information and was able to act had the driver accidently fallen out of the train or became ill during the journey. This true story explains why Health and Safety is so important and why it must be a legal process and policed by the Health and Safety Executive. Thankfully, the company changed management several years ago and now takes safety very seriously and conducts full induction programmes. The driver was perhaps equally responsible if not more, because he failed to consider his obligations towards passenger safety. This is why certain laws are passed, because left to their own devices companies, and indeed their employees, would not necessary face up to their responsibilities.

Learning outcome 3.3

Identify the most appropriate communication methods for use in work situations.

Communicating effectively is an art in itself, and involves three distinct stages:

1. Speak to a person
2. Person accepts information
3. Check for understanding

This third stage is vital – you must confirm that the person understands what has been asked of them. This is not necessary an easy task because we actually communicate through many different formats such as:

- Talking directly to people
- Holding meetings
- Scanning technical drawings
- Email
- Telephone conversations
- Phone, including mobiles
- Skype
- Computer conferencing
- Letters

In order to create the correct impression, it is important to use the correct format for different occasions. For instance, when applying for a job or communicating with a client, for instance, then communication should be made through a formal, properly drafted letter. There are, however, different ways of addressing the

Description

Communication is a three-stage process.

Description

When communication breaks down, 99% of the time the blame lies in the transmitting end. This is because they should always check that the receiving end has understood the message.

person you are writing to. For instance, if you are known to the person, then use the greeting:

Dear Mr Roberts

and finish the letter by signing off with 'yours sincerely'.

When making contact with a person unknown to you use:

Dear Sir/Madam

and finish the letter by signing off with 'yours faithfully'.

During certain occasions there is a need to pass information on to a large number of people and conducting a formal meeting is very effective in this regard. This is because meetings can permit an opportunity for those present to discuss any concerns, issues or problems they are currently experiencing. It also provides time to highlight what are termed priority tasks, in other words, it will pinpoint completion of tasks that are regarded as the most important and the most urgent. Quite often the need to prioritise is borne out of certain delays that have stemmed from unforeseen problems such as:

- Experiencing severe weather
- Waiting for materials to be delivered
- Staff illnesses
- Unanticipated structural damage
- Serious accident and follow up investigation

It is important to stress that when creating a new schedule it should be workable and conducive to all the trades involved, including alerting any trade not present at the meeting of all relevant details. A bar chart, sometimes referred to as a Gantt chart (Figure 4.63 a & b), is a type of management technique used to plot the various stages of a development. Certain stages or operations are then planned to be completed by specific dates, which means the Gantt chart can highlight if the development is on track or lagging behind schedule.

Time / Activity		Day number													
	1	2	3	4	5	6	7	8	9	10	11	12	13	14
A	░	░												
B	░	░	░	░	░	░	░	░						
C			░	░	░	░								
D							░	░	░	░	░	░	░	░
E	░	░	░	░										
F	░	░	░	░	░	░								
G		░	░	░	░	░	░	░	░					
H						░	░	░	░	░	░			
I		░	░	░	░	░	░	░						
J	░	░	░	░										
K							░	░	░	░	░	░		
L										░	░	░	░	
M													░	░

A simple bar chart or schedule of work

Figure 4.63a

Continued

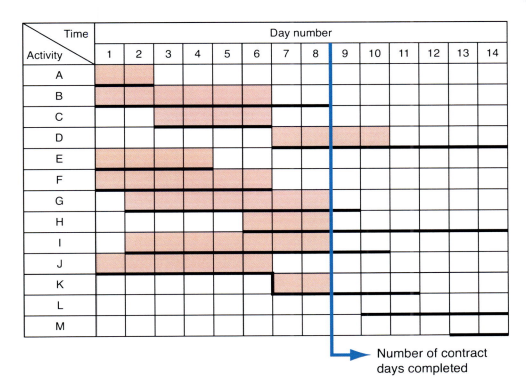

Number of contract days completed

A modified bar chart indicating actual work completed

Figure 4.63b

Figure 4.64 t is vital that there is a written record of verbal agreements affecting final plans

Far less public and intimate conversations also happen on a construction site and can also be an effective way of discussing issues, raising concerns or giving instructions. However, when verbal agreements are made between two different people, especially if they affect any final plans, then it is vital that a written record is made in order to record the details of what was settled and arranged (Figure 4.64).

Top tip

Always write down verbal agreements. Some people have selected hearing and memory function.

Top tip

Never disclose another person's personal details.

Language and tone

A very important consideration of any communication process is to alter your language accordingly when speaking to different people. Whilst you should always be respectful irrespective of whomever you speak with, your manner and tone can be strikingly different. For instance, two apprentices chatting to each other would be classified as informal. In comparison, if the same apprentices were speaking to a client, architect or a Health and Safety Inspector, their tone should be very different. Remember, in many circumstances you are effectively representing your company and you must take care in how you respond. Equally important, never offer advice or answer questions unless you are fully aware of the facts involved. If in doubt, it is far better and wiser to refer the issue to your immediate supervisor.

Despite the fact that communication takes place all the time, it is incredibly important to understand that personal information should never be divulged. This is why the Data Protection Act 1998 was devised and passed as law. This ensures that a person's personal details are not disclosed without their permission.

For example, consider the following situation.

Scenario A

An employee answers the company telephone and during the conversation gives the caller their employer's personal mobile phone and home address. Has this situation broken the Data Protection Act?

Scenario B

An employee answers the company telephone and during the conversation refers the caller to the company website and address of the company warehouse. Has this situation broken the Data Protection Act?

Scenario A has violated the Data Protection Act, because the employee has passed on their employers personal details without their permission. Scenario B however has not, because information in relation to a company website and warehouse are already in the public domain.

Written communication

There are a variety of different written communication formats used in the electrical installation industry, each with its own purpose. For instance, when a customer requests some work, the details are recorded on a job sheet (Figure 4.65), which not only tells the electrician what the job entails but will indicate what kind of equipment and materials they require.

Time sheets (Figure 4.66), on the other hand, record the time spent on each job and are then used to prepare weekly or monthly pay slips. Be aware, failure to complete a timesheet could mean that the person involved will not be paid at the end of the week. Both the job sheet and time sheet are then used to work out the costs involved in a particular job, which is then transferred on to the customer's bill.

Occasionally, a customer will ask for extra work to be carried out, over and above what was originally agreed. Day work sheets are therefore used to record the time and materials used to complete this additional work.

A very handy form of written communication is known as a memorandum, or memo, and is especially convenient when recording telephone messages.

Description

A job sheet indicates what kind of wiring system is required.

Top tip

Always fill in your timesheet, otherwise you might not be paid.

```
┌─────────────────────────────────────────────┐
│  JOB SHEET            SPARKS                  │
│  Job Number ........  ELECTRICAL             │
├─────────────────────────────────────────────┤
│                                               │
│  Customer name  ............................. │
│                                               │
│  Address of job  ............................ │
│                                               │
│                 ............................. │
│                                               │
│                 ............................. │
│                                               │
│  Contact telephone No.  ..................... │
├─────────────────────────────────────────────┤
│  Work to be carried out  .................... │
│                                               │
│  ............................................ │
│                                               │
│  ............................................ │
│                                               │
│  ............................................ │
├─────────────────────────────────────────────┤
│    Any special instructions/conditions/materials used │
└─────────────────────────────────────────────┘
```

Figure 4.65 A job sheet

For instance, it can outline the time and date of the call, who the caller is and what exactly is being asked or requested. I know from personal experience of what it is like to forget to pass on a message and having to face up to the consequences; therefore a memo is extremely useful in this regard. An example is shown in Figure 4.67.

It is also good practice to do an inventory check of any materials when they are delivered on site. This can be done by comparing the delivery note, against a purchase order, in order to ensure that the materials ordered is the same as that delivered.

The use of email is also effective for many different reasons; for instance, email attachments are able to transfer huge amounts of data to a variety of different people. Email systems also store any incoming or outgoing information in its history. This is why many people use it to send information to themselves, using their email facility as a memory drive. Email is also very effective if you need to raise important points or safety concerns, because you can send an email directly to your immediate supervisor. However, if you felt that these safety issues have not been addressed and continue to pose a danger to employees. Then you can forward your original email up the management chain, with your sent history retaining a copy of the original complaint.

It is important to highlight that for communication to be effective your message must be clear to all those involved. For instance, using text talk in emails might confuse certain people or will result in the receiver not taking your

Top tip

Always spell check emails or any other word-processed work.

TIME SHEET

SPARKS ELECTRICAL

Employee's name (Print) ---

Week ending --

Day	Job number and/or address	Start time	Finish time	Total hours	Travel time	Expenses
Monday						
Tuesday						
Wednesday						
Thursday						
Friday						
Saturday						
Sunday						

Employee's signature --- Date --------------------

Figure 4.66 A time sheet

SPARKS ELECTRICAL

internal **MEMO**

From _____ *Dave Twem* _____ To _____ *John Gall* _____

Subject _____ *Power Tool* _____ Date _____ *Thrus 11 Aug. 13* _____

Message

Have today ordered Hammer Drill from P.S. Electrical – should be with you end of next week – Hope this is OK.   Dave.

Figure 4.67 An internal memo

message seriously. Care should be taken not to include any use of swear words, which is not acceptable or appropriate in any form of communication.

Lastly, there is another form of communication which is still in use today and is known as a fax, or facsimile to give it its proper name. When sending a fax you are sending an exact copy of a letter or other document through a machine. Not many companies rely on this form of communication, because to work properly you require two machines, one at the transmitting end and one at the receiving end. To receive a copy both machines need to be on, but unlike email there is no real means of storing any sent history and the quality of any images reproduced can be very poor.

Functional and technical information

Electrical installation also makes use of both functional and technical information.

Functional information is data written in such a way that explain how something works or operate, and is typically found within manufacturer's instructions. Contrast this to technical information, which allows you to compare measurements to see if they meet a particular specification or test.

Technical information for an electrician comes in many forms and would include:

* Safe Isolation Procedure (GS 38)
* Wiring Regulations (BS 7671 & On-Site Guide)
* Manufacturer's instructions and manuals
* British Standards (BS) and British Standard European Norms (BS EN)
* Technical books and drawings
* Company websites and forums

The Wiring Regulations (BS 7671) bring about best practice for electricians because they impose certain core values but also standardised a range of different information. For instance, all symbols have to be drawn from BS EN 60617, which will safeguard that designers or installers of an installation in Britain and Europe can readily identify with any of the components stated.

There are other sources of information available, such as manufacturer's instructions which are extremely useful and informative when installing equipment. A manufacturer's manual, on the other hand, will indicate how any equipment is meant to be operated, including its various operational settings and controls. I have personally witnessed many a fault being reported on an aircraft but on closer inspection it became apparent that the fault disappeared if the maintenance procedure was followed correctly.

Important point

Certain technical documents such as the Wiring Regulations are designed to provide best practice.

Internet

The Internet is a vast store of information that anyone with access can contribute to. Problems can occur, however, if information is blindly extracted and used in a technical context. For example, I once read a post on an online forum which openly stated that a 6 mm cable is sufficiently large to supply any type of shower outlet. The problem lies in the fact that you cannot verify the person's qualifications, or credentials, which means that you cannot verify the credibility of the information.

Any qualified electrician worth their salt will tell you that you should not generalise what size cable should be fitted to electrical equipment. There is no one size fits all category in the Wiring Regulations. The size of cable needed will

Top tip

Always challenge the credibility of information.

be determined by the power rating of the shower, value of the supply voltage, its means of routing and protection as well as in certain circumstances requiring correction factors to be applied if its environment contains excessively high temperatures or comes into contact with thermal insulation. Furthermore, certain equipment such as cookers are permitted an allowance called diversity which can reduce the size of the cable required, but this does not apply to showers.

An extra appendix (Appendix C: Credibility of information) covering the importance of establishing the credibility of information has been included at the end of this book.

Learning outcome 3.4

Identify the actions to take when conflicts arise.

Learning outcome 3.5

State the effects that poor communication may have in the workplace.

The effects of poor communication can seriously weaken the on-going success of a business or undermine a building project. Lack of progress and failure to meet deadlines are often associated with low morale within the workforce when the employees do not buy in to the aims and objectives of their employer. This could well stem from the fact that certain levels of management are failing to keep their workforce informed and they in turn feel isolated.

When key stages in a building programme are not met, not only can penalty fines be imposed but it can also affect the reputation of that company. This is potentially far more damaging, because although a good reputation can take years to build up it's possible to damage it in seconds. It can also affect the confidence that a client or building contractor has in that contractor and given that they pay the bills, meeting their approval is vital for any future business.

Important point

Communication should be a constant process:

- Employers talking to employees
- Employees talking to employers.

Figure 4.68 *Arguments need to be settled sooner rather than later*

It is critical, therefore, that communication lines remain effective, all technical trades should be kept informed of their priorities and be allowed opportunity to raise issues and concerns so that they can be resolved. This is known as working for a common cause or meeting organisational goals. If employees are kept in the loop then they feel involved and their work is valued, which in turn will maintain their enthusiasm and motivation. Remember, a happy workforce is an effective workforce.

Conflicts between workers and management or different trades do occur (Figure 4.68), but they must be resolved immediately by arranging a meeting with all the affected parties. Unfortunately, disciplinary procedures is sometimes necessary and are used against employees with regard to:

- Bad time keeping
- Improper behaviour
- Substandard workmanship

Substandard workmanship however could have resulted from the employee not being given proper supervision which is a statutory requirement through the Electricity at Work Regulation 1989. In such circumstances, if an employee is a member of a trade union, then such an organisation would challenge their dismissal on the basis of a failure to fulfil employer obligation and unfair treatment. A trade union could also be approached if an employee was subjected to certain rituals such as being sent on humiliating errands to fetch:

- A bubble for the level
- A bag of sparks for the grinder
- A long weight
- A glass hammer
- Elbow grease
- Some tartan paint

Although this type of practical joke is portrayed as a bit of fun, it alienates and belittles the individual concerned and should be seen as a form of bullying and challenged and outlawed accordingly.

Human factors

When an accident happens, the subsequent investigation will normally uncover one common cause: a breakdown in communication. Historically, accident investigations tended to blame one person, without looking beyond and examining any other evidence or related facts. The importance of communication and the link to industrial accidents is so important that a whole new area of research has been developed, and it is known as 'human factors'.

When an accident investigation looks at the 'human factor', this means looking at the accident from every perspective in order to uncover not only why it happened but more importantly, how it could have been avoided. A human factor investigation will therefore look at:

- What training was given to workers?
- Was there any pressure exerted?
- Were there any distractions?
- Was the accident caused by bad or uncorrected habits?
- Was there a lack of supervision?
- Were the policies or procedures at fault?
- Were the correct or approved tools used?
- Why were certain checks not carried out?
- Were the correct materials used?
- Was the test equipment calibrated?
- Were the installers working shifts?

Top tip

Any disagreements on site should be resolved as soon as possible.

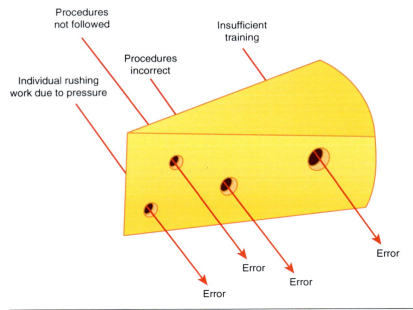

Figure 4.69 Errors can slip through certain holes in procedures

A very important consideration for electrical installation apprentices is the following statement: Some of the best people make the worst mistakes.

This was examined earlier in Chapter 1 through the tragic story of how Ken Woodward lost his sight. Ken, to his credit, tried to find a solution to a problem, but never considered his own safety. There have been occasions when supervisors have exerted pressure on their workers but on this occasion the pressure was self-imposed. Ken embraced a 'can-do attitude' but should have thought 'can I do it safely?'

Another consideration for electrical installers is that faults do not always show themselves straight away. This type of fault is known as a latent fault, and will perhaps take days, weeks or even longer before it manifests itself. This is certainly the case in Australia, whereby substandard cable has been installed in 40,000 domestic properties and the situation is basically a ticking bomb.

Both are examples of how error can get through certain holes in procedures as indicated through the cheese slice diagram Figure 4.69.

The holes could be caused by:

- Insufficient training
- Training or good practice not always used
- An individual being stressed
- Incorrect procedures
- Procedures not always followed (takes longer if you do it correctly)
- Individuals who dismiss their training or do not have a positive attitude

The permit to work (PTW) system has already been described as one mechanism that brings about a safe system of work; essentially, it is a controlled document that authorizes and details the requirements and lists any special precautions and procedures to be followed. The PTW will be raised by the 'authorized person', who will detail the requirements and specify that certain safety measures are in place, then sign/date the document. The contractor will read the document, sign/date it and abide by it when carrying out the work. When the work is complete, the 'permit to work' document is signed off (work completed, system safe to work on).

There are two famous examples of how a safe system of work broke down, which led to the loss of life. Both incidents include some element involving electrical workers.

The first is Piper Alpha, which was an oil platform that was destroyed on 6 July 1988, when an initial oil fire caused high pressure gas lines to explode, killing 167 people. It is still the worst offshore oil field disaster to date in terms of human life lost. The main cause of the disaster surrounded reinstating a pump that was offline for maintenance. The engineers believed that the pump had only been electrically disconnected, but in fact a blanking plate had been fitted instead of a pressure valve. When they tried to operate it, the blanking plate failed and an explosion occurred. The work had been authorized on two permits, one for the maintenance and another regarding the fitment of the blanking plate. The permit for the removal of the pressure valve was lost and not taken into account. This was most definitely NOT a safe working environment.

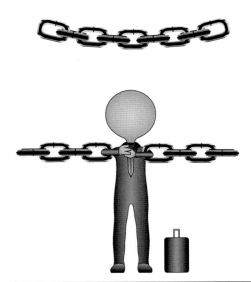

Figure 4.70 It only takes one person to break the accident chain

The second incident is the Clapham Junction rail crash, which ironically occurred only five months after Piper Alpha. The Clapham Junction rail crash was a multiple train collision that occurred at roughly 08:10 on 12 December 1988. The enquiry found that a wiring fault had stopped the alarm system from working and therefore alerting an approaching train that another train lay ahead. It became apparent that a certain technician had never been corrected regarding some of his bad practices such as leaving defective and redundant cables in a loom. Even worse, the technician responsible did not insulate the ends of the redundant cables which then shorted out the alarm involved. The work of the technician was never checked; therefore, a clear lack of supervision was also responsible for this disaster. Bad working practices, lack of supervision: again, definitely NOT a safe working environment.

When accidents happen, it is usually a chain of events. A single individual can break the chain: the incident or accident will not happen. Put simply, Health and Safety regulations, examples of good practice and codes of practice are designed for a reason. Always follow them, speak up when they are not in place and speak up if you do not have the correct equipment or materials. Be that person (Figure 4.70) and break the chain (Figure 4.71).

'Human factors' reminds us that electricians should use certain 'Codes of Practice' such as referring to the Wiring Regulations which are written so that any

Figure 4.71

statutory requirements are met. It is also vital that safety features brought out by the Electricity at Work Regulations 1989, especially with regard to adopting safe isolation procedures, are used. Safe systems of work are maintained by adopting risk assessments and applying method statements and permit to work systems. But systems are only as good as the people who use them; therefore, always adopt a positive approach to safety.

Time out, let's summarise some **key** points!

- A successful installation relies on a team of people, comprising of many different trades, and each assigned a specific role
- Each person has a duty of care not just to their employer, but to their co-workers including being mindful of their responsibilities towards: timekeeping, dress code and on-site behaviour
- Management should keep team members informed of what is going on, especially outlining which tasks should take priority
- Work schedules are designed to plot and monitor progress, but following any unexpected delays, they would need to be re-evaluated by all the trades affected
- When conflicts occur then they need to be dealt with as soon as possible

To review your understanding, try this True or False quiz based on communication.

Question 1: An architect produces a set of plans and technical drawings.

True	False

Question 2: When deliveries arrive on site, they should always be checked against a purchase order.

True	False

Question 3: Most accidents involve a breakdown of communication.

True	False

Question 4: Data protection means that you are allowed to give away personal information with consent.

True	False

Wiring systems and enclosures

Match the meaning of the following by placing the appropriate number next to the letter in the box provided (the first one has been done for you).

Thermoplastic PVC sheathed and insulated cable (flat profile) A	A = 6	Cable type with very high level of mechanical protection and is often installed underground as a sub-main (1)
Flexible conduit B	B =	Cable type used in high temperature and explosive environments (2)
Mineral insulated (Pyro) C	C =	This type of wiring enclosure is used as the final connection to a motor circuit (3)
PVC/SWA D	D =	This type of wiring enclosure would need to be warmed up to room temperature before being used if left outside overnight (4)
Bus-bar system E	E =	Power system used in industrial environments that require high current use (5)
Plastic conduit F	F =	Cable type used in domestic properties (6)

Drills and fixings

Match the meaning of the following by placing the appropriate number next to the letter in the box provided (the first one has been done for you).

| HSS drill bit A | A = 3 | Type of fixing used in hollow walls such as plasterboard (1) |

| Masonry drill bit B | B = | Heavy duty type of fixing used to secure items in stone (2) |

| Spade drill bit C | C = | Type of drill bit that should be used with metal (3) |

| Spring toggle D | D = | Type of drill bit that should be used with stonework (4) |

| Rawlbolt E | E = | Type of drill bit that creates holes in wood (5) |

| Plastic plug F | F = | Type of fixing used with brick (6) |

Components

Match the meaning of the following by placing the appropriate number next to the letter in the box provided (the first one has been done for you).

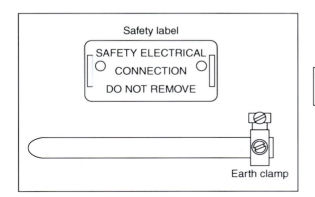

A = 4

Ceiling rose – loop-in wiring system
(1)

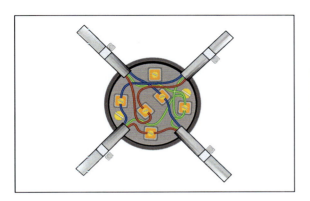

B =

Earthing sleeving required for twin and earth cable
(2)

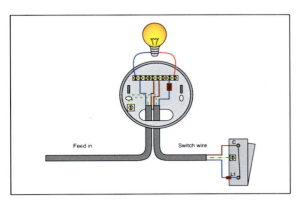

C =

Junction box
(3)

D =

Clamp – British Standard BS 951
(4)

Electrical installation methods word search

Find the following hidden words:

EXTRANEOUS, EXPOSED, CHALKLINE, HSS, TORQUE, HIGHRESISTANCE,
SPRINGTOGGLE, RAWLBOLT, SWA, FLATPROFILE, RCD, METALCONDUIT,
FUSE, CIRCUITBREAKER, LOOPIN, CFL, LED, RING, RADIAL, EARTH, TRAY,
LADDER, IR, JOBSHEET

E	X	P	O	S	E	D	L	I	T	Q	R	P	G	R	U	H	E
X	R	E	A	N	O	N	H	M	E	M	I	O	N	E	M	I	O
T	O	R	Q	U	E	L	E	E	E	E	N	W	I	K	S	G	H
R	Q	C	L	A	I	D	A	R	H	C	G	E	K	A	N	H	M
A	A	D	C	W	O	O	E	A	S	I	E	A	N	E	M	R	N
N	K	D	O	E	D	E	L	E	B	Q	H	R	U	R	I	E	E
E	T	L	O	B	L	W	A	R	O	E	R	T	R	B	N	S	N
O	Y	A	R	T	H	C	T	I	J	S	K	H	T	T	O	I	I
U	G	A	T	L	O	O	P	I	N	R	I	N	C	I	R	S	L
S	L	F	L	A	T	P	R	O	F	I	L	E	R	U	C	T	K
W	E	S	I	S	T	A	N	C	E	D	A	E	E	C	I	A	L
S	P	R	I	N	G	T	O	G	G	L	E	S	D	R	M	N	A
A	S	T	E	R	N	A	T	O	R	G	F	U	D	I	R	C	H
A	M	H	E	C	F	L	R	E	M	R	O	F	A	C	A	E	C
M	E	T	A	L	C	O	N	D	U	I	T	R	L	I	A	W	S

Test your knowledge

1. Know the key tools and fixings used in electrical installation.

1. What type of tool is used to remove sharp bits of metal from conduit?
   a. wrench
   b. knife
   c. reamer
   d. screwdriver

2. An electrician's screw driver should be:
   a. coloured red
   b. made from aluminium
   c. made from plastic
   d. insulated

3. Small hand tools can be protected from rust by:
   a. applying some varnish
   b. applying some paint
   c. covering with a light coating of oil
   d. washing it often

4. To mark out metal trunking you would use:
   a. pencil, tape, spirit level
   b. steel rule, square, scriber
   c. hacksaw, file, centre punch
   d. plumb line, square, spirit level

5. Which of the tools listed is best suited for clipping cables into place?
   a. mallet
   b. square
   c. hammer
   d. chisel

6. Which is most accurate?
   a. tape measure
   b. laser line levels
   c. bubble level
   d. plumb bob

7. Which of the terms below is used to indicate if a surface is level?
   a. out of line
   b. in-line
   c. bubble line
   d. plumb

8. All tools and equipment should be checked:
   a. before use
   b. after use
   c. at the end of the day
   d. before and after use

2. Know the basic requirements of wiring support systems.

9. Domestic properties use what kind of cable?
   a. metal conduit
   b. twin and earth
   c. mineral insulated
   d. PVC non-sheathed single cables

10. The main supply that is fed into a house is sometimes referred to as a sub-main. What kind of cable is associated with this term?
    a. SWA
    b. fibre optic
    c. MI
    d. PVC flat profile

11. The term strapper is used with which kind of lighting control system?
    a. radial control
    b. three-way lighting
    c. one-way lighting
    d. two-way lighting

12. Why does a modern house include a separate ring for the kitchen?
    a. it is normally warmer
    b. it is normally bigger
    c. it normally contains a lot of electrical equipment
    d. it normally contains a cooker

13. Domestic properties are normally fitted with what kind of circuit breaker?
    a. type A
    b. type B
    c. type C
    d. type D

14. What kind of lighting control is used for a room having only one entrance?
    a. pull switch control
    b. intermediate switch control
    c. one-way switch control
    d. two-way switch control

15. What kind of lighting control is used for a room having two entrances?
    a. pull switch control
    b. intermediate switch control
    c. one-way switch control
    d. two-way switch control

16. An electrical cable is made up of three parts, which are known as:
    a. outer conduction, inner sheath and radiation
    b. conductor, insulation and outer sheath
    c. conductor, outer insulator and inner sheath
    d. conductors, insulators and sheaths

17. A fuse can be considered as:
    a. the strongest part of the circuit
    b. the weakest part of the circuit
    c. the highest part of the circuit
    d. the lowest part of the circuit

18. In a domestic property most of the protective devices are housed in a:
    a. three-face distribution board
    b. plastic box
    c. plastic enclosure
    d. consumer unit

19. A power cable that does **NOT** loop back to its supply circuit breaker is called a:
    a. radial
    b. ring
    c. wring
    d. roundabout

20. The plug top below is fitted with what kind of cable?
    a. SWA
    b. fibre optic
    c. PVC flex
    d. PVC flat profile

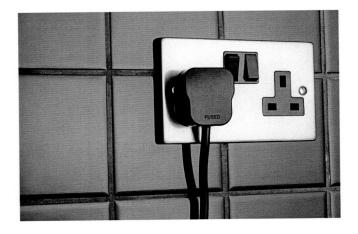

21. The purpose of a cable sheath is:
    a. to provide its insulation
    b. to provide some form of colour
    c. to provide protection from impact damage
    d. to provide flexibility

22. What kind of situation had occurred in the image below?
    a. overheating
    b. underheating
    c. chemical effect
    d. magnetic effect

23. What kind of component is shown below?
    a. junction box
    b. ceiling rose
    c. mains adapter
    d. loop in box

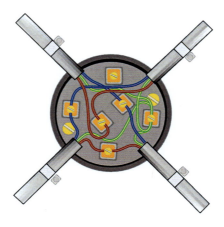

24. Which of the following is a document that an electrician would use at work?
    a. On-Site Guide
    b. internet
    c. Wikipedia
    d. microfiche

25. Ladder racking is a system for:
    a. supporting PVC sheathed cable
    b. supporting BT cable
    c. supporting armoured cables
    d. supporting metal conduit

26. Which of the following would be best to provide the final connection to an electric motor?
    a. PVC conduit
    b. plastic conduit
    c. metal conduit
    d. flexible conduit

27. A fuse protects:
    a. the person
    b. the switch
    c. the cable
    d. load

28. An RCD protects:
    a. the person
    b. the switch
    c. the cable
    d. load

29. All electrical connections should be:
    a. fairly low in resistance
    b. fairly high in resistance
    c. high in resistance
    d. very low in resistance

30. Earthing is a system that ensures a circuit is isolated when a fault occurs. But which **TWO** of the following can achieve this?
    a. transformer
    b. generator
    c. fuse
    d. circuit breaker

31. Very large loads such as cookers, showers and large water heaters should be:
    a. supplied through its own dedicated ring circuit
    b. supplied through its own dedicated circuit
    c. supplied through a battery backup
    d. supplied through either a ring or radial circuit

32. Circuits are divided and organised to ensure that:
    a. no single fault affects all circuits
    b. a single fault affects all circuits
    c. a single fault only affects some circuits
    d. no single fault trips all circuits

33. Which **TWO** of the following environments would be suitable for steel conduit?
    a. classrooms
    b. domestic property
    c. farm
    d. industrial

34. Which **TWO** of the following environments would be suitable for plastic trunking and conduit?
    a. classrooms
    b. domestic property
    c. commercial
    d. industrial

35. Within a consumer unit, the highest rated circuit breaker should be positioned:
    a. next to the main switch
    b. opposite end to the main switch
    c. next to the main RCD
    d. next to the second RCD

36. When wiring up a consumer unit, the downstairs lights should be positioned alongside:
    a. the upstairs lights
    b. the upstairs sockets
    c. the downstairs sockets
    d. the kitchen ring circuit

37. Which of the following will **NOT** operate during an overload but provides additional protection?
    a. transformer
    b. fuse
    c. RCD
    d. circuit breaker

3. Know the requirements of working with others in electrical installation.

38. If one particular trade fell behind another then it is best:
    a. to punish all the trades involved
    b. to ignore the trade involved
    c. to complain about the trade involved
    d. to help the trade involved

39. It is recommended that verbal agreements:
    a. are written down
    b. can only take place between two people
    c. must always involve an apprentice
    d. are photographed

40. Communication is a:
    a. one-stage process
    b. two-stage process
    c. three-stage process
    d. four-stage process

41. Which of the following is most hazardous?
    a. installing bigger cables than required
    b. installing smaller fuses than required
    c. installing equipment in the wrong place
    d. installing equipment not fit for purpose

42. If bad weather affects a construction project, which of the following would be the best course of action?
    a. hope that everything will work out
    b. recruit more staff just for this job
    c. reschedule any work activities affected
    d. pray for better weather

43. To promote good customer relations an electrician or an apprentice should always:
    a. ignore clients
    b. speak respectively to clients
    c. refuse to speak to clients
    d. refuse to acknowledge clients

44. When an electrician or an apprentice is asked their opinion, they should only answer if they:
    a. know a little about the situation
    b. know everything about the situation
    c. know something about the situation
    d. think they know about the situation

45. When a contractor is first meeting up with a client on an official basis, this should be thought of as:
    a. a formal meeting
    b. a casual meeting
    c. an informal meeting
    d. a relaxed meeting

46. Informal methods of communication are suitable when:
    a. attending an interview
    b. attending a disciplinary procedure
    c. making first contact with a client
    d. discussing something with a colleague

Unit: ELEC1/05A

Chapter 4 checklist

Learning outcome	Assessment criteria – the learner can:	Page number
1. Know the key tools and fixings used in electrical installation.	1.1 Identify the key hand tools and their uses. 1.2 Identify maintenance requirements for hand tools. 1.3 Recognise power tools and accessories. 1.4 Identify types of fixings.	138 138 143 145
2. Know the basic requirements of wiring support systems.	2.1 Identify the components of wiring systems. 2.2 Recognise the application of wiring system components. 2.4 State the items to look for when checking finished work. 2.5 Recognise documents which affect installation of wiring systems.	149 149 174 176
3 Know the requirements of working with others in electrical installation.	3.1 Recognise the main roles of the site team. 3.2 State the importance of company policies and procedures that affect working relationships. 3.3 Identify the most appropriate communication methods for use in work situations. 3.4 Identify the actions to take when conflicts arise. 3.5 State the effects that poor communication may have in the workplace.	178 181 183 190 190

EAL Unit: QSWC1/01

Starting work in construction

Learning outcomes

The learner will:

1. Know about different types of career opportunities available in construction.
2. Know about different types of organisations offering career opportunities in construction.
3. Understand how career choices can impact upon an individual's lifestyle.
4. Be able to make informed career choices.
5. Be able to work responsibly with others.
6. Be able to seek and respond to guidance when working as part of a team.

EAL Electrical Installation Work – Level 1. 978-1-138-23206-8
© 2017 P. Roberts. Published by Taylor & Francis. All rights reserved.

Learning outcome 1.1

Describe different types of career opportunities in construction.

Professional and technical careers

Electrical installation fits into a distinctive part of the construction industry and is involved in installing electrical systems within three key environments: domestic (housing), commercial (offices, schools, classrooms, hotels) and industrial (factories, refineries, etc.). The construction industry is not only an important industry but is a major employer and contains a range of different companies. For instance, there are major building contractors, medium to small companies as well as individual traders. A sizeable development such as that awarded to erect and build a new school is normally given to a large building contractor, who will employ other companies as sub-contractors to carry out specialist work. Such specialist work includes electrical installation, plumbing, joinery and roofing. But the good news is that not only do these firms provide a service but they can also provide career opportunities.

My father once told me that there are two types of employee:

- Those who shower before they come to work
- Those who shower after they finish work

You cannot apply these distinctions today, because manufacturing and mining industries such as coal, steel and slate are vastly reduced and have all but disappeared in many locations. But as an ex-slate miner, he distinguished between those who worked inside an office and those situated outside – regardless of the weather. These office jobs are sometimes referred to as white collar, but in truth and in fairness to the people involved, many occupations are recognised and given professional status, but need certain academic qualifications to degree level and above to achieve this. It is true to say that the construction industry relies on both professional and craft occupations to operate successfully. The importance of this relationship is shown in Figure 5.1.

Any job role will include both soft and hard skills but what exactly is the difference? Well, for an electrician, hard skills include the work involved in designing and installing an installation, whilst soft skills involve the ability to communicate and work effectively with other people such as your work colleagues and other skilled workers. This is why they are sometimes referred to as interpersonal skills. Every profession in the construction industry uses and applies these skills; therefore let us examine each occupation in turn, whilst also focusing on how they can provide career opportunities.

An **architect** is given professional status if they are affiliated with the Chartered Association of Building Engineers, a leading body for professionals specialising in the design and construction, evaluation and maintenance of buildings. An architect can be thought of as an interpreter since they are expected to produce working drawings, which reflect the needs and requirements of their client. It is vital, therefore, that an architect understands the variety of ways visual

Tackle a difficult word: interpersonal skills

Ability to communicate and interact with other people

Description

An architect talks with their client and then produces a set of technical plans which match their requirements.

Figure 5.1 A successful building project requires both professional and craft operatives

information can be represented. For instance, isometric drawings show two-dimensional images of three-dimensional (3D) objects.

Because plans and drawings need to be extremely detailed (Figure 5.2), architectural technology such as CAD computer software has been developed which produces images and graphics in 3D. An architect will also supervise certain aspects of the build to ensure their plans are interpreted correctly, therefore good communication skills are essential. Another job role, referred to as a **Clerk of Works**, assists an architect on site by making regular checks.

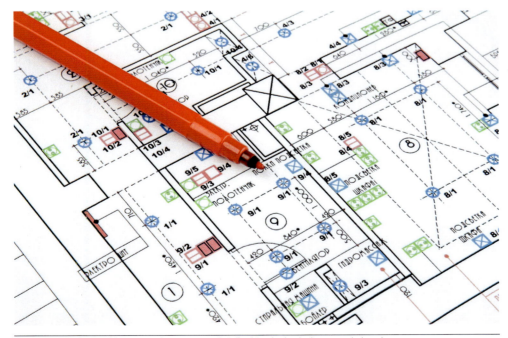

Figure 5.2 An architect produces very detailed technical plans and drawings

A clerk of works acts as the eyes and ears of the architect to ensure that the project is progressing as intended. This job role offers someone with the relevant construction experience a different type of career as a liaison person between the professional and the skilled workforce.

The Chartered Association of Building Engineers also include **surveyors**, who make a living looking at and examining buildings and any land that it is built on, and use specialist levelling equipment such as an infrared reflector as shown in Figure 5.3. A very important part of a surveyor's role is to offer advice to clients about the design and construction of a new building, including, where appropriate, any care, repair and maintenance required.

Figure 5.3 A surveyor uses high-quality land surveying equipment to gauge the lie of the land

This work is closely associated with that of **quantity surveyors** who are concerned with all the materials and labour costs involved in the project. In fact, all costs as well as any profit margins would have been submitted through what is known as a tendering process. This is when companies submit sealed bids that register their interest in carrying out certain work for a stated sum of money. If you have an interest in buildings or a curiosity about materials and costings and possess good communication skills then this could well be an occupation for you.

Very large engineering and transport projects are designed and run by **civil engineers**; for instance, the construction of modern schools, sky scrapers and even bridges. In addition, modern buildings are designed to be largely self-sufficient by the installation of alternative technologies such as solar PV and wind turbines, which will generate sufficient electricity for its use. If you are interested in designing and actually managing how things are built then this job could be ideal for you.

Understanding how material and structures withstand load or heavy weight is the responsibility of **structural engineers**. The term strength actually means an ability to withstand heavy loads and structural engineers need to understand how

Quantity surveyors

Calculate all the costing involved in an installation including all the labour and materials required.

Description

A compression force applies pressure or force, which can squeeze, squash or compact an object. Putting sleeving on electrical cables is a good example.

different forces – such as shear, tension and compression –affect materials and buildings.

It might not surprise you to learn that engineers need good communication skills, but equally they require an excellent understanding of mathematics. Some of the calculations involved are very difficult and are consequently carried out by computers; therefore, it goes without saying, familiarity with computer software is also essential. Often, structural engineers are also asked to investigate problems and they need an excellent grasp of scientific knowledge in order to solve problematic situations.

Craft occupations also need to understand what is meant by the term 'load-bearing structure'. For instance, this term applies to particular load-bearing walls that are responsible for shouldering the weight of a building. This is in direct contrast to the installation of partitions, which simply divide up the interior space of a property, but they themselves are not load-bearing. Does being a structural engineer appeal to you?

A new role has become fashionable; if you want to work in an area with a growing emphasis on the importance of reducing the impact that buildings and people have on the environment, then perhaps you might want to look for a career as a **building services engineer**. They look after and monitor different environmental systems such as electricity demand, control of heating and waste processes in order to reduce the amount of greenhouse gases produced. Because of the variety of systems involved, it is not unusual for building services engineers to have entered this field already qualified in other areas such as electrical engineering/installation, heating or any other appropriate degree. An example of a modern eco-friendly building is shown in Figure 5.4.

There are also various different management careers possible within the construction industry. The term management actually means controlling or organising something, which could involve people, materials or job roles. **Construction management**, better known as **project management**, is carried out by the person who plans, coordinates and controls a project from beginning to completion. Project managers can either look after several small projects or a

Description
A tensile force can be thought of as what happens during a tug of war, when power is exerted through the rope, as it is pulled tight from opposite ends.

Description
A shearing force pushes one part of a body in one direction and another in the opposite part of the body. A good example is when electrical cable is cut.

Description
Building services aims to reduce the impact on the environment by making buildings as energy efficient as possible.

Tackle a difficult word: optimal
The **best** solution

? Did you know?
Being objective means looking at other perspectives or opinions and not just your own.

Subjective means looking at something through your eyes only.

Figure 5.4 Building services engineers control heating and minimise electricity use to reduce the impact that buildings and people have on the environment

single large development. They need to possess certain skills, such as keeping control of finances and costing, ensuring all work activities remain on schedule, and communicating with other agencies and professions. A project manager also needs to apply critical thinking – the ability to weigh up a situation and look for solutions in an objective manner.

Once machines or buildings are in use, they need to be maintained, which allows career opportunities in **maintenance management**. Maintenance managers need more than good technical knowledge of the equipment/buildings they manage, because they also need to manage a team of people alongside reviewing all procedures so that all machinery and general operations operate as efficiently and as safely as possible. They will also ensure that certain maintenance operations are carried out during what are known as the quiet hours, such as after midnight or even at weekends, so that productivity is not affected. This process is known as time management. In many industrial environments, some machines include components that are categorised as critical and are so important and vital to operations that every now and again they are replaced even if they remain in good working order. This is because their failure can lead to serious and grave consequences. Replacing components before they fail is known as preventative maintenance, but replacing one that fails unexpectedly and prematurely is identified as corrective maintenance. It is during these types of situations that problem solving skills come into their own; a person needs to look at every option and resource available in order to devise an optimal solution.

A manager's communication skills should always include being approachable and adopting an open door policy, in order to encourage employees to truthfully report problems and mistakes. Without confidence in their manager, employees are more likely to ignore difficulties or equipment malfunction, which can lead to much more serious situations.

Another possible career lies in the field of **facilities management** (Figure 5.5). The manager makes sure that buildings and whatever business or service they provide match the needs of the people who work, visit or use them. A facilities

Believe it or not

Maintenance managers will use charts that are shaped like a bathtub in order to try and predict component failure.

Figure 5.5 Facilities managers match the needs of their customers and provide them with the facilities required

manager requires very good communication skills because they need to maintain different working relationships when informing and advising an array of different people of the availability of various services. This position requires certain management and organisational skills such as allocating key roles to specific staff members to ensure that the customer's or client's needs are met. Does this sound like you?

Craft and operative careers

If you are not drawn to these types of professional occupations, then perhaps you might consider a career as a skilled manual worker in a particular trade or craft. Such work is associated with a famous expression: 'being on the tools'. Craft type careers generally serve an apprenticeship, which typically can range from three to five years.

Certain people prefer to be physically engaged, or involved, in a task rather than any deal with any academic content. In other words, they do not want to talk about something, they want to get on and do something! However, no craft occupation is without their theoretical element, because all trade-based activities are underpinned by their respective mathematical and scientific content.

The difference is that the mathematical content is rooted within the theoretical knowledge and this is sometimes referred to as 'applied mathematics'. For instance, working out how much carpet is required to cover a given floor space involves applying the formula for area = length x width (m^2). The price of carpets tends to be sold in square metres, but value-added tax (VAT) is also added to each purchase and is normally shown as a percentage (%). When the customer receives the bill, it will include the following elements:

- Purchase price of the carpet
- VAT (20% of cost price)
- Cost of fitting carpet (£17/hour, the hourly rate)

Calculating volume on the other hand involves three-dimensional shapes: Volume = length x width x height (m^3).

Although the formula remains the same, the term changes to Capacity when calculating the storage of fluids or gases but is measured in litres (1 m^3 = 1 Kilo litres or 1000 litres).

Other examples of applied mathematics include the use of ratios. For instance, when a plasterer makes a cement mix, a typical ratio would be four buckets of sand to one bucket of cement. The size of an electrical cable is actually a measure of its cross-sectional area, which alongside its length and conductor material, will determine the resistance of the circuit. An electrician needs to understand these types of calculations because it is part of the design process in defining what size cable is required in a particular circuit. If such calculations are applied incorrectly, then at best the circuit will fail to operate, but at worst it can lead to an electrical fire or electrocution!

Choice of craft career

The choice of which craft occupation to follow as a career preference should be driven by what actually interest you. For instance, a **bricklayer** is involved in establishing the structure of a building either through brick or block. On small projects, bricklayers can also get involved in creating the actual foundations of a

? Did you know?

VAT is a type of consumption tax that is added to the cost of a product.

building, which is normally a solid concrete base, and is built up using blockwork to ground level. That said, on larger projects, this kind of work of preparing and installing the foundation stage of a building is carried out by a team of people known as groundwork staff.

An important consideration of most craft based occupations is that they can be labour intensive. Being a brickie or roofer, for instance, can literally be back breaking work but equally very rewarding as you witness the building being physically assembled. Another potentially arduous job is that of a **stonemason**, who is involved in repairing stone on old buildings, or working with stone blocks on modern new build houses. A stonemason also erects stone walls on public roads, either by using mortar or through dry stone walling techniques, in which walls are assembled by selecting and meshing various types and size of stones. Lastly, some stonemasons are commissioned to create sculptures and various types of memorials.

Carpenters, also referred to as joiners or chippies, are actually involved in several areas of construction, ranging from installing wooden door frames, skirting, sills, partitions and wooden trusses and joists. Joiners also have to apply certain elements of mathematics not only when measuring and fitting but also in design. For instance, wooden joists used in the support of a building are actually rated to withstand different weights known as loads. This means that when installing joists in an attic that contains several water tanks, the joists would have to be strong enough to support the weight of the header tanks because water, when supplied in large volumes, is very heavy. Shuttering joiners, on the other hand, use what is known as formwork or wooden planks of ply to make temporary moulds to tip concrete. Carpenters can also specialise in producing objects such as wooden tables and chairs. Even early boat building relied on an ability and understanding of how to manipulate specialised craft techniques on wood or wood-based products.

In the construction industry, certain skills are classified as wet trades, which means that they use materials that have to be mixed with water. **Plasterers** are a prime example, as are bricklayers, given that they both use water to create mortar. Plasterers usually apply three layers to a surface – the initial layer is called a scratch coat and the others gradually build upon this layer (Figure 5.6). Some plasters use pink gypsum to create a final finish. Plasterers apply their trade to both inside as well as outside, coating the exterior of a building with pebbledash unless the building is a brick finish wall, but other forms of rendering are also possible.

Sometimes certain crafts include two distinct but different skills such as **painting** and **decorating**. Trainees are again expected to serve an apprenticeship in order to learn their craft and, trust me, wall papering is an art in itself. In addition, it takes time to fully understand how different varieties of paint and varnish are applied to different materials and surfaces. 'In vogue' is a term that is used to indicate a prevailing fashion and trends do come and go. For instance, up to the 1990s, most houses were decorated with wall paper, but modern housing tends to be adorned with decorative paint. Who knows what will happen in the future?

Occasionally, materials used in construction are adapted and sometimes the introduction of cheaper alternatives has a direct bearing on any craft trades involved in this sector. For instance, the introduction of tiled roofs meant that **roofers** had to adapt their skills base beyond slate, which had been the predominant method of roofing for many years. Although a specialist alternative, thatched roofs are also still in existence but are normally installed by specialist

Figure 5.6 Various trades have to work together, such as an electrician and a plasterer

contractors. However, given the limited demand for straw roofs, the skills and trade secrets involved could eventually die out and be lost forever, unless further generations of thatched roofers enter the trade and maintain and resume this distinctive and unique craft.

Figure 5.7 Electricians tend to work in three specific areas: domestic, commercial and industrial. They can also specialise in just one

Another two trades which are closely associated are **plumbing** and **electrical** installation, since they are sometimes collectively known as 'building services'. This is because they both provide services and facilities such as gas, water and electricity. Plumbers tend to install everything involving water, such as equipment found in bathrooms, kitchens and waste water systems. Some plumbers broaden their qualifications and are then authorised to install and maintain gas-operated equipment such as boilers, cookers and gas fires. **Gas fitters** also maintain and install such equipment, but all craftspersons who undergo such work must be registered with Gas Safe, a regulatory body that ensures that the skills and competency of all those registered is maintained and assessed.

Of all the trades listed, an electrician offers more transferability than most, since the term electrotechnical spans everything from someone who fixes fridges to someone who works on motorway street lighting. Some specialise in just one aspect, such as a domestic installer, also known as a 'house basher'. Alternatively, an industrial electrician is someone who will only install or maintain large-scale industrial equipment (Figure 5.7). Certain electricians are also classified as maintenance electricians, not necessarily installing equipment but carrying out both corrective and preventative maintenance on various types of equipment.

Kitchens and bathrooms offer another avenue of employment regarding **wall** and **floor tiling** specialists, who earn a living not only in domestic properties but also kitchens and toilets in public and commercial locations such as pubs and restaurants. Similar to wall papering, it is an art to match up the patterns involved and therefore a creative eye might be beneficial.

All the craft trades described are skilled operatives but there are unskilled roles known as general construction operatives, sometimes referred to as labourers

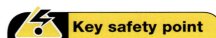

Key safety point

Only qualified and Gas Safe registered employees can legally work on any gas-operated equipment.

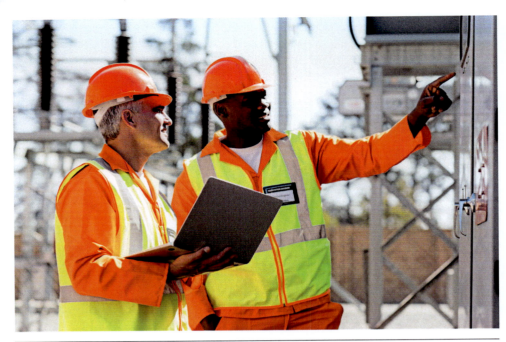

Figure 5.8 General construction operatives require supervision at all times

or even a plumber's or electrician's mate. Although not necessarily qualified, this is not to say that they are without any skill or understanding but, in general, they do not possess the depth of knowledge of a qualified trade person. This means that they should be supervised at all times (Figure 5.8). Both professional and craft occupations all combine to create the building team as shown in Figure 5.9.

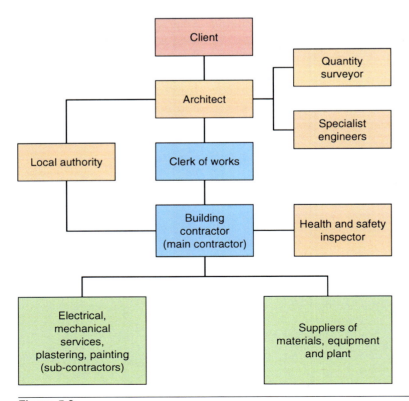

Figure 5.9

Occupation task activity

This activity contains two sets of matching terms and each matching pair should be given the same number. The first one has been done for you: 'Electrician = 1' matches up with 'Installing cables = 1'

Electrician = 1	Plumber = 2	Installs cables = 1	Installs pipes =
Works with wood products =	Works with plaster =	Painter and decorator =	Matches building to the requirements of different businesses = 8
Architect = 5	Quantity surveyor = 6	Plasterer = 3	Joiner = 4
Works with paints, varnish and wall paper = 7	Facilities manager =	Produces detailed plans and drawings =	Works out material costs =

Type of employment

When people seek employment they normally contact an employer directly or make contact through a family friend. Alternatively, job centres and on-line agencies advertise job vacancies including apprenticeships. Depending on the person and their needs or skills, they could be drawn to seek full-time employment or can either find or be offered a part-time job.

A contract of employment can be either permanent or temporary, in which case it will come to an end or is renegotiated after a specific amount of time.

The hours involved in a full-time contract can vary, but will normally be around 40 hours a week, presuming that you start at 08:00 and finish at 16:00. It is important that such details are made clear to you before you start your position, but, more important, you need to read and understand before you accept and sign any contract of employment. A contract of employment will normally include:

Important point

Always read a contract of employment in full before you sign it.

- Hourly rate of pay, including any overtime or bonus pay
- Hours of work, including any expectation that you may have to work overtime, night shifts, weekends and bank holidays
- Paid holiday entitlement
- Sick pay
- Redundancy pay
- How much warning you need to give the employer before you leave
- How much notice an employer must give you if you are made redundant

As stated previously, part-time employment can also include a contract of employment, adjusted accordingly, but might not include such things as holiday pay and sick pay. It is also worth stating that some people only gain employment during certain parts of the year, especially for those people who are employed in the hospitality or tourist sector and gain employment during the summer months. This is often referred to as seasonal work.

There is an alternative, called self-employed status, where a person effectively works for themselves and runs their own business. This appeals to some people because you are effectively your own boss. That said, certain self-employed trades are taken on by a main building contractor and are therefore referred to as sub-contractors. Occasionally, a client will identify a preferred contractor to carry out a given service, which then makes that particular firm a nominated sub-contractor.

One last point but a very important one concerning craft based trades – contractors by definition go from contract to contract and when the going is good and work is abundant they tend to take on and employ quite a few people. Unfortunately, there is a flip side because when such work or contracts come to an end, employers can also lay people off.

Learning outcome 2.1

Describe different types of organisations that offer career opportunities in construction in terms of their size and the nature of the work they undertake.

As described previously, there are various ways of obtaining employment, but there is a subtle difference between a career and a job. Although these two terms can amount to the same thing, a job is mostly concerned about earning some money, but a career is a choice of occupation such as a long-term pursuit of a lifelong ambition.

There are many different types of organisations in the UK, with the following types of businesses relevant to construction and electrical installation. Multinational companies are very large organisations that manage production or deliver services in more than one country. This means that although some multinational organisations will base their headquarters in one country, they assemble or produce a product in other countries. Coca-Cola, for instance, is based in America but distributes its soft-drink range around the world. In the construction industry, Balfour Beatty has interests in many building projects worldwide.

There are national companies which operate solely in one country such as Wales, although sometimes that can extend to the other home nations. Gilks Nantwich Ltd., for instance, is an electrical installation firm that operates in Nantwich (England) and on RAF Valley (North Wales). Equally Barratt and Taylor Wimpey are two of the UK's most successful construction and development companies and the top two house builders. These types of large organisations tend to invest and get involved with large public or government works or construct extensive large-scale housing estates or new school and college projects.

Large multinational construction firms also require a range of different materials. Large building materials suppliers such as Jewson, Wicks and Aggregate, to name just a few, provide this service, but this means they can also provide career opportunities.

In the electrotechnical industry, other companies exist in the supply and distribution of electricity such as Scottish Power and First Hydro who are themselves owned by a joint venture between International Power Ltd. and Mitsui & Co. Ltd.

In contrast, there are small and medium enterprises, more commonly called SMEs, which are much smaller in size and whose employee numbers and annual turnover fall below certain limits. SMEs are normally involved in repair and maintenance, extensions and minor building works. Quite often, however, these smaller companies, who specialise in one or more areas of expertise, such as electrical installation, will be invited to tender to a main contractor to carry out all the necessary electrical work.

What all these companies have in common is that they offer apprenticeships and other types of employment. Most companies have a company website, which includes very useful information about their history and core business as well as interests in other sectors.

Learning outcome 3.1

Explain how an individual's lifestyle may be influenced by the career choices they make.

Before embarking on any career, it's important to carry out research in order to collect as much information as possible, including gathering the opinions and testimonies from other people such as family and friends. Sometimes people are persuaded to enter an occupation because they have heard of the money they can earn. However, please consider that it will take a lot of time, effort and hard work for you to climb up the ladder and be paid the same as a person who has been involved in that industry for a number of years.

When considering various careers consider:

What qualifications do you need for a specific route?

Some companies/colleges require a certain number and grade of GCSE, but others will recruit after an interview process and scrutinising the results of their own aptitude tests. Before embarking on any career path it's essential that you enquire what qualifications are required.

Craft based occupations involve serving an apprenticeship, which means that you must secure full time employment with an appropriate employer of your chosen trade.

Although further education colleges do offer a range of craft courses to full-time learners on completion of the course, they too must find an appropriate employer if they wish to progress and qualify as a skilled operative.

The training and development that will be needed.

Certain careers such as electrical installation require that you attend and complete a Level 3 combined award which consists of theoretical examinations as well as a portfolio of evidence to prove that you can use certain skills in a real working environment. The electrotechnology industry also expects practicing electricians to maintain and top up their skills and keep abreast of any new

Important point

To be signed up as an apprentice, you must first secure work through an employer. You cannot fully qualify as a craft specialist by attending a full-time course.

changes or technologies that are being brought into the industry. This is known as CPD (Continual Professional Development).

Their general state of health.

This is a very important consideration, because certain manual work is very demanding on the body. Asthma sufferers are not normally allowed to join the Armed Forces, unless they can prove that they have not been prescribed medication for four years. Eye sight problems and even colour blindness will also restrict what trade routes they can follow. That said, it is important to remember that certain laws have been put in place to stop unscrupulous employers discriminating against prospective employees and include:

- Equal opportunities
- Sex discrimination

The conditions under which they will work.

Most construction trades including electricians will need to at some point:

- Work at height
- Work in confined spaces
- Work in hazardous environments
- Work outside in wet and cold weather

The possible effect on their personal relationships.

Some employment is contractual, that is they work on one project and hopefully move on to another and so on. This can mean, however, that you have to:

- Work away from home
- Work extended days
- Work a night shift
- Work during weekends
- Be on call or standby

The general demands of the chosen career.

We have already discussed that if you intend to follow a craft based occupation, there are certain physical demands, the ability to use and apply mathematics as well as reaching a certain level of craftsmanship. The good news is that all this can be learnt, but you must be prepared to work hard and embrace a positive attitude to learning.

Important point

Always consider any health implications before deciding on a career choice.

Important point

Claustrophobia is the irrational fear of confined spaces.

Acrophobia is the irrational fear of working at heights.

Most craft based trades will at one time work in confined spaces and at height.

Learning outcome 4.1

Make realistic career choices based upon information provided.

Training and development requirements

This was considered previously but is incredibly important when reviewing any career choice. Many companies will ask for three GCSE's such as English, mathematics and a science-based subject. They are sometimes referred to as prerequisites, but for those individuals who have not achieved such qualifications, they are left with little choice but to consider other types of employment. There is one alternative, however, which is to consider finding a school or college in order to acquire them. Some colleges and learning providers offer night classes and even online courses, but always obtain proper advice to ensure that they are reputable (trustworthy).

Organisation types

Think and research what options are available to you in your local area, especially ones that interest you. For instance:

- Are you aware of any large or medium contractors in your local area (or even beyond if you drive and possess your own transport, or are lucky enough to have reliable public transport links)?
- Have you actually carried out any research such as looking in the local paper, Internet, local radio or even job centres and employment agencies?
- Do you have a family member or friend that is currently working in a job or industry that interests you? If you do, speak to them personally or email them for advice.
- Have you thought of writing and asking a company for some work experience? If you make a good impression, arrive to work on time and demonstrate a positive attitude, if a vacancy or opening arises the position could be yours.

Types of Activities

The field of 'electrotechnology' is vast and relates to any use of science that deals with the practical application of electricity. This means that some contractors install electrical installation systems including, in some instances, re-wiring caravans. Others maintain electrical systems such as those in factories, street lighting and even overhauling motors, generators and refrigerators. The Armed Forces maintain aircraft, ships and vehicles including tanks. All of these are applications of electrotechnology, so research what is available to you but, more importantly, research what interests you!

Progression

Companies may have various and different progression routes, for instance many managers were once apprentices, who started at the bottom rung of the ladder but have worked their way up. Electrotechnology more than any other trade has what is known as 'high transferability'. I know it sounds like a made up term, but it's true! It relates to identifying a set of skills in a particular environment and transferring them to a different setting. I qualified as a domestic electrician whilst I was still in the RAF even though I was working mainly as an aeronautical electrical engineer. When I left the service, I was employed as a civilian Training Officer, interviewing potential candidates who had applied for conversion training to become aircraft electricians. They included an ex-soldier from the Armoured Corp (Tank) Regiment, an electronics engineer and a car mechanic. Their previous qualifications and hand-skills were recognised which boosted their credibility as applicants and ultimately they were seen as being transferable. Skills are not enough, however; they also needed to work extremely hard and thankfully they did and were successful in completing their conversion training and embarked on a new career within the aeronautical industry. Many employers will value and recognise qualifications you have achieved including Level 1 courses and this will hopefully allow you to continue to further and greater things.

Did you know? ❓

Most employers value communication skills and a positive work ethic above all else.

Transferable skills

Employers everywhere tend to value the same core skills. It is true to say that they will often look for certain grades of GCSE but other employers tend to value 'attitude before aptitude', in other words they value soft skills more than hard. This is because if you possess the correct attitude then most technical skills can be learnt, but attitude comes from within a person and is not so easy to learn, apply or put in place. There are always difficult situations to overcome, problems to solve and customers to please. This is why good interpersonal

Important point

Transferability is very important. It means certain core skills can be recognised in other industries.

skills and organisational ability are probably the most important abilities you can offer, alongside an enthusiastic nature that is willing to put the team before themselves. Some core skills are listed below:

- Punctuality and good time management
- Attention to detail
- Ability and willingness to ask for help when required
- Willingness to accept responsibility
- Self-awareness: knowing your strengths – and weaknesses
- Literacy and numeracy including mental mathematics
- Commitment and motivation
- Interpersonal skills: can you relate with others and form good working relationships?
- Communicating effectively: can you both take and give instruction?

Benefits

It is important to list the benefits of any prospective career and the following could influence your choice:

- Pay (earnings)
- Travel
- Transferable skills
- Non-contributable pension (you do not have to pay towards a pension)
- Secure employment
- Company operate locally
- Use of company vehicle
- Healthcare schemes

Expectations

It is important to establish what your expectations and perhaps answering certain questions might help.

- To what lengths are you willing to go to find a new job or career?

- What kind of sacrifices are you willing to make?

- Are you willing to travel? Do you want to travel?

- Do you have the commitment to serve a four-year apprenticeship?

- Are you aware that the minimal wage does not necessary apply to an apprentice?

Learning outcome 5.1

Demonstrate good team working skills by working responsibly and cooperatively.

All companies and industries rely on effective communication. It matters not if you work in retail, hospitality, tourism or heavy industry, it is crucial that individuals work well with others in a team environment, because for any organisation to be successful it needs cooperation and harmony. In other words employees need to pull together to meet all organisational goals and targets. For any team to be effective it needs all the individuals involved to think of themselves in a wider context and embrace a sense of collective responsibility. In other words, the team or task comes before any of the individuals or personalities involved.

There are two further important considerations when working in a team environment. Although it is important that all those involved give of their best, if any individual appears to be struggling or not contributing, it might be that they are unsure of what to do or they are not comfortable with their role. It is far better to offer assistance rather than shout abuse.

On the other hand, it is vital that if you as an individual are struggling or are not sure what it is you are meant to be doing, then be honest and ask for help and guidance. It takes maturity to be the bigger person and openly admit that you need help.

Important point

There is no **I** in team.

Top tip

A team is only as strong as its weakest link.

Learning outcome 6.1

Follow instructions when working with others.

The definition of an active listener will be covered in Chapter 6 but adopting some of its core principles is incredibly important when following instructions. To listen effectively an individual must listen carefully to what is being said, especially if the person is emphasising or repeating any particular aspect.

I must admit at such times in the past, due to my shocking short term memory, I made a point of making some notes in relation to the instructions given. I found that this technique not only refreshed my memory when needed, but boosted my confidence, which in turn allowed me to contribute meaningfully to the task ahead.

Engineering procedures on the other hand, never rely on the memory of their operators, this is because memories fade and engineering practices change. No electrician including the author can recall everything that is contained in the Wiring Regulations, but we can adopt techniques that teach us how to find information, such as searching out key terms in the index.

There are many examples of engineering faults being reported not because any of the equipment involved was faulty, but rather because the person who has set up the equipment in the first place had not read the manufacturer's manual or user guide properly. Being able to follow instructions correctly is not just very important but **incredibly** important, because distractions can occur all too easily and our minds can wander at the best of times.

To counter this, I always read instructions at least twice, to ensure that I fully understand what it is I need to do and in what particular order. This is also very important when you sit exams; read the question very slowly so that you understand EXACTLY what is being asked.

Giving instructions

Before the arrival of satellite technology, I have on numerous occasions parked up in a side street and got out to ask a passer-by where exactly I am, and in which direction should I travel in order to reach my destination. I have been given some strange directions over the years, such as:

'I wouldn't start from here!'

Or,

'At the end of the road there is a right turn, after that there is a pub called the White Horse, don't go that way!'

Sometimes in trying to be as helpful as possible, people give too much detail such as:

'Go to the end of the road, left at the roundabout, right at the T junction, then left again at the Garage. Continue for about 5 miles, then right at the traffic lights, this will take you through an industrial estate. When you come out, go left and up the hill, over a very sharp bridge and you will spot a football field. The building you want is behind the cricket pavilion.'

The point that I am trying to make is that during such occasions I should have asked them to repeat the instructions or write them down. Instead, I simply nodded my head and politely said thank you and ended up having to ask several more people before I arrived two hours after my estimated time of arrival.

Golden rule: If you are ever given some instructions, or directions for that matter, and you are unsure of what was requested or said, ask the person to repeat it! Better still, write it down!

Classroom activity

Many famous people started life off as electrical apprentices or indeed in some cases completed degrees in electrical engineering. This is because young adults in particular are advised to gain a trade so that they have something to fall

---

Did you know?

It's difficult to listen to another person without interrupting. This is why it is called the art of listening!

back on if their preferred career does not pan out as they hope. They include footballers, musicians, politicians and actors.

Why not research some of the following names to see if this applied to them?

- Stuart Pearce ('Psycho')
- Rowan Atkinson
- David Jason
- Bobby Charlton
- George Harrison
- Elvis Presley
- Lech Walesa

Craft or professional route activity

To help you consider what kind of occupation best suits you, put an X against a box which interests you most, but also make a note of its number. Once you have done this, use the number to reflect against a second table shown at the end of the chapter. This will then tell you the craft or professional occupation you have chosen.

Craft		Professional status	
1. Studying through a further education college		1a. Studying through a university to gain a degree	
2. Working with your hands		2a. Working with materials and installation costs	
3. Installing cables		3a. Producing detailed construction drawings and plans	
4. Maintaining systems		4a. Ensure that buildings match different business requirements	
5. Working with bricks or blocks		5a. Working with CAD	
6. Working with mortar (plaster)		6a. Working with engineering load baring structures	
7. Working with wood materials		7a. Working with environmental building controls	
8. Creating stone structures		8a. Managing projects, people or facilities	
9. Roofing buildings using slate, tiles or even straw		9a. Working with the construction of large scale engineering projects such roads and bridges	

Research

Once you have established or expressed an interest in a particular occupation, carry out further research with the companies listed below and any others that you can find online. Look at their location, what they do, but more importantly – what can they offer you?

Construction Industry Training Board
http://www.citb.co.uk/

Balfour Beatty
http://www.balfourbeatty.com/

National Grid
http://www2.nationalgrid.com/uk/

Bardsley Construction
http://www.bardsley.co.uk/

Lovell
http://www.lovell.co.uk/

Carillion
https://www.carillionplc.com/

Laing O'Rouke
http://www.laingorourke.com/

Morgan Sindle
http://www.morgansindall.com/

Kier
http://www.kier.co.uk/

Mitie
http://www.mitie.com/

Interserve
http://www.mitie.com/

The Construction Index
http://www.theconstructionindex.co.uk/

Unit: QSWC1/01

Chapter 5 checklist

Learning outcome	Assessment criteria – the learner can:	Page number
1. Know about different types of career opportunities available in construction.	1.1 Describe different types of career opportunities in construction.	208
2. Know about different types of organisations offering career opportunities in construction.	2.1 Describe different types of organisations that offer career opportunities in construction in terms of their size and the nature of the work they undertake.	218
3. Understand how career choices can impact upon an individual's lifestyle.	3.1 Explain how an individual's lifestyle may be influenced by the career choices they make.	219
4. Be able to make informed career choices.	4.1 Make realistic career choices based upon information provided.	220
5. Be able to work responsibly with others.	5.1 Demonstrate good team working skills by working responsibly and cooperatively.	223
6. Be able to seek and respond to guidance when working as part of a team.	6.1 Follow instructions when working with others. 6.2 Communicate appropriately with others.	223

Answers to craft or professional route activity

Craft	Professional status	
1. Apprenticeship	1a. Graduate	
2. Working with your hands	2a. Quantity surveyor	
3. Installation electrician	3a. Architect	
4. Maintenance electrician	4a. Facilities manager	
5. Bricklayer	5a. Design software	
6. Plasterer	6a. Structural engineer	
7. Joiner	7a. Building services engineer	
8. Stonemason	8a. Project manager	
9. Roofer	9a. Civil engineer	

EAL Unit: QMSA1/01

Managing study and approaches to learning

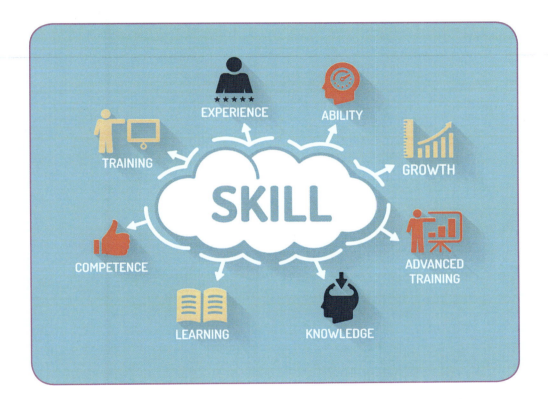

Learning outcomes

The learner will:

1. Understand the demands of a course of study.
2. Understand how to organise study time effectively.
3. Understand how to prioritise and set realistic targets for study.
4. Be able to find and use information relevant to the course of study.
5. Understand how to listen in and contribute actively to a learning environment.
6. Keep information in a usable format.

EAL Electrical Installation Work – Level 1. 978-1-138-23206-8

Learning outcome 1.1

Identify the main aim of a course of study.

Learning outcome 1.2

Outline the demands of a course of study including: timescale, attendance, assessment criteria, forms of assessment and self study.

If you are reading this particular chapter then hopefully it's because you have aspirations to become an electrician. It is true to say that an electrician is a unique trade – not only do they handle and manipulate electrical cables but they are also deeply involved in fabricating systems made out of metal and plastic. They also need certain skills normally associated with other trades such as:

- Brickwork – plastering over chases in walls
- Woodwork – removing floor boards or drilling through wooden joists

Given these facts, it is very important that, when choosing a college or learning provider, prospective students ensure that they are equipped with proper facilities and are accredited to industry recognised awarding bodies such as EAL. It is equally important to ensure that as a potential applicant you meet any course requirements, which are sometimes referred to as **prerequisites**, which list any qualifications or GCSE grades required for entry.

All students who enrol on a course are asked to sign a learning agreement, outlining certain expectations and behaviours that all learners are expected to abide to. This mirrors similar procedures that are used in industry, to highlight that all employees have a duty of care not only to their employer but also to other employees.

It's equally important that students consider if they can meet the commitment and demands of the course by establishing:

- How long the course lasts
- How many days a week attendance is required
- Teaching start and finish times

Meeting this commitment is incredibly important, since it is a proven fact that excellent attendance is related to success.

It is also a good idea to research how the course is assessed. For example, exams can be internally marked or assessed online. Practical assessment can also differ, with some centres continually assessing their students throughout the academic year whilst others insist on an end of course practical test. Some courses are taught online, through what is sometimes referred to as distance learning, or through postal correspondence but it has to be said such means would not be fit for purpose as a way of qualifying for a craft based trade. This is why as mentioned earlier it is very important that you ensure that your chosen course and learning provider can provide you with a proper pathway that suits you and your chosen occupation.

Important point

It's vital that you find out what is involved and required to enrol on to a course. If you are accepted then you must make a commitment to see it through.

Important point

You have got to be in it to win it. Excellent attendance is the key to success.

Be aware

Not all types of calculators are allowed in examinations.

Learning outcome 2.1

Identify appropriate times of study.

Learning outcome 2.2

List the key features of a good, safe and productive learning environment.

Learning outcome 2.3

Identify the personal challenges that may affect study.

For any learning to be effective it needs to be relevant to an individual, because one person's method of study will not necessarily work for another. For instance, some people prefer to study with music blaring out in the background, while others need peace and quiet so that they can focus on their reading or a particular activity. It is very important, therefore, that you discover what kind of environment works for you, but firstly you need to establish:

- The date of the examination
- The time of the examination
- Where is the examination being held

It is equally important to find out if any specific equipment is required for a particular exam, such as a programmable calculator. However, be aware that some scientific calculators that include a means of storing formulas are not permitted.

Some exams are referred to as open book, which means that the exam questions have been written and drawn from one or more specific text books and each candidate is required to bring a copy to the examination. Forgetting to turn up with the required books means that you stand little or no chance of success. Given that many colleges now charge a resit fee, forgetting to organise and prepare for an examination could even lead to extra expense.

Well in advance of the exam, therefore, there is a need to plan when and where you are going to study and consider how to establish a good, safe and productive learning environment. Evidence suggests that studying at the same time and same place every day is effective for most students. What is definitely **not** effective is ignoring the issue and hoping for the best. Even those students whose only strategy is to stay up the night before an examination, trying to cram in as much information as possible will nose-dive at some point or other.

Think of it like this: If you fail to plan, plan to fail.

Planning study skills session is even more important when there are other challenges to overcome. For instance, some people hold down part-time or even full-time jobs. Other people, including in some cases young children, are actually carers and have responsibility towards other family members. Dyslexia and other specific learning disabilities can also add to the stress involved when sitting examinations. If you have been formally assessed regarding dyslexia or other specific learning difficulties, then please ensure you make the college and your tutor aware. This is important because in certain circumstances and dependant on your assessed needs, extra help is permitted, such as being given extra time for the examination or you can be allocated a reader – a person who reads out the questions on behalf of the candidate.

Important point

Don't turn into an ostrich and stick your head into a pile of sand. Find out when and where your exam is taking place.

Important point

Plan your study programme.

How, when, where?

The why is obvious – studying leads to exam success!

Important point

Dyslexia should be thought of as a learning difference. You think differently to other people and it is not related to intelligence. If you have been diagnosed with dyslexia or suspect that you learn differently or are experiencing difficulties with learning, then alert your tutor.

Exam stress

From a physical point of view, some people think better after exercising, as an active body leads to an active mind. I must stress that this does not apply to everyone, but even if you are not in this way inclined, certain relaxation exercises might help you de-stress before you start to study and before an exam.

Mindfulness is a type of meditation technique which tries to focus your mind on external elements such as sounds and sensations. I must admit I have tried to meditate in the past and focus on what is called and known as the third eye; in other words, having greater control over your mind and emotions. But it never worked until I tried some mindfulness techniques. I am not an expert and I would recommend that if you are interested in this subject then you attend a course. It is said that when people they lose their sight, their hearing improves. There is no actual medical evidence to prove this, but I can believe that being forced to explore their sense of hearing would enhance their sharpness and sensitivity. I certainly found that when I was particularly anxious, and after ensuring that it was safe to do so, closing my eyes and trying to focus and identify all the different background sounds around me helped me move on from whatever was troubling me.

What does affect everyone, although to different degrees, is a lack of sleep. Examinations can be stressful, but if you are tired as well then your concentration levels are going to be severely tested. Alcohol has a bearing on this, if you are old enough to drink, of course. This is because alcohol stops your brain from attaining deep sleep, which is why although it might make you drowsy and tired at first, a lot of people wake up in the middle of the night, unable to sleep any further.

> **Important point**
>
> Try to relax before an exam, and research some relaxation methods such as mindfulness techniques.

> **Did you know?** **?**
>
> Depriving someone of sleep has been used as a form of torture, so go to bed early the night before your exam.

Study periods

It is believed that it is far better for the average person to study over a lot of small periods rather than one long period. Quality tends to be better than quantity because your initial focus and concentration at the beginning of a session will be at its sharpest, but this will gradually decline over time. Taking regular breaks will enable you to restart on a high again, maximising the amount of information you take in.

Learning outcome 3.1

Outline what SMART targets are.

Learning outcome 3.2

Identify SMART targets for own study.

SMART is a very helpful acronym which can be used to design an effective study programme. It stands for:

S – specific

M – measurable

A – achievable, acceptable, action-oriented

R – realistic

T – time-based or time-bound

Tackle a difficult word: acronym

An abbreviation formed from the first letter of a group of words

For example:

Specific – Set a programme of studying two hours per night for a week.

Measureable – The sessions will be measured by completing fragments of study.

Achievable – Is this study programme achievable?

Realistic – Is this study programme realistic?

Time-bound – Run the programme over a week.

You need to be honest and true to yourself when setting these type of targets and whilst they do not always work because other commitments get in the way or you experience countless complications, most difficulties can be overcome. The word prioritising means arranging items in order of importance and, at exam time, finding time to study should be high on your to-do list. Therefore, if your initial study programme is not working, re-evaluate by applying SMART and re-think your study strategy. By all means change and adjust the programme, but it is essential that you embrace a 'can do' attitude.

Important point

Can you succeed?

Yes you can!

Adopt a 'can do' attitude!

Learning outcome 4.1

Use a range of reference systems to locate specific information.

When looking through books, especially technical books, both the contents and index pages are extremely useful in order to find any relevant information. The contents pages, with a typical arrangement shown in Figure 6.1, will break the book down into chapters, but more importantly, these chapters are tied to page numbers. The index (Figure 6.2), in my opinion, however, is actually more useful because all key words within a specific chapter are listed in alphabetical order, alongside its associated page number. If the index lists several pages, then

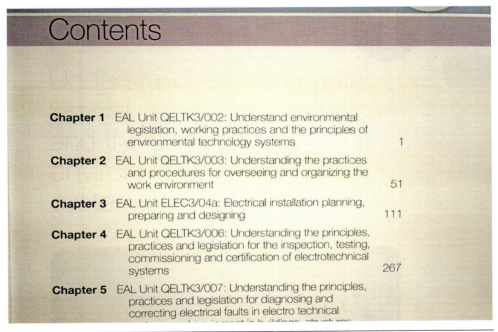

Contents

Figure 6.1 The contents page breaks the book down into chapters and gives their page numbers

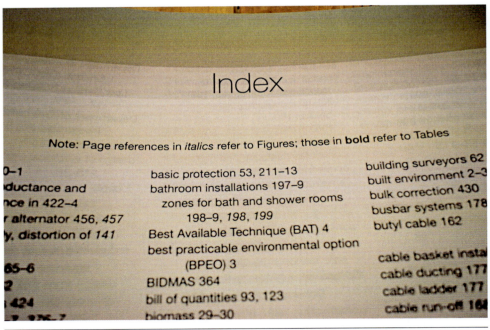

Figure 6.2 The index of a book will highlight key words in alphabetical order as well as their corresponding page number

simply write them all down, so that you can look at each in turn until you find what you are looking for.

University students are encouraged to do this and is often referred to as 'mugging books' because they are not expected to read all their journals, books and Internet resources in their entirety, but simply extract any relevant information.

There are other techniques that can help; for instance, the pages in some books are colour coded so that each chapter stands out. You can also ask your teacher or tutor, to summarise which pages you need to focus on if you are weak on any particular aspect. EAL produce a post-exam printout, and although it will not contain any of the examination questions, it will identify which learning outcomes require further study and consideration. Ask your tutor for a copy.

Last, many technical books will include useful summaries in an appendix located at the back of their book including a glossary of terms so that you can review any difficult or technical words as well as a summary of important formula.

Internet

The internet is a huge library of information but 'surfing the internet' can be tricky, because search engines will often return millions of hits. However, there is a simple trick you can use to reduce the number of results returned and it involves putting speech marks ' ' around the words entered.

When speech marks are used, the search engine is basically filtered to only recognise and return those sites that include all the words typed in. A word of warning, this technique is how teachers can tell if a piece of written work submitted by a student has been lifted straight from the internet. All the same, it is a very effective technique and useful way of defining your search parameter.

Top tip

When researching books, the index at the back of the book will often list terms in alphabetical order – very handy.

Top tip

Narrow down your internet research by using speech marks: 'Electrical installation jobs'.

Use a range of reading techniques for different purposes.

There are four basic reading techniques a student can use when reading or studying and each one offers something different and will be more appropriate to different types of reading.

They are known as:

- Skimming
- Scanning
- Intensive
- Extensive

The process of reading is actually very interesting; for instance, can you make sense of the following passage?

Aoccdrnig to a rscheearch at Cmabrigde Uinervtisy, it deosn;t mttaer in what oredr the ltteers in a word are, the olny iprmoetnt tihng is that the frist and lsat ltteer are at the rghit pclae. The rset can be a total mses and youcan still raed it wouthit porbelm. This is bcuseae the huamn mnid deos not raed ervey iteter by istlef, but the word as a wlohe.

Most people, unless you suffer from a specific learning difference such as dyslexia, can read the article without any great difficulty. This is because a lot of the words are fairly common and we automatically recognise them as long as the first and last letters are in place. This type of ability allows us to skim and scan words but let us look a little deeper into the difference between the two.

Skimming

This is very useful if you are looking for a specific word or principle such as 'earthing', for instance, but is not very good for actually understanding its comprehension or understanding what context it has been used.

Task

Skim the following passage and decide how many times the letter F appears.

Finished files are the result of years of scientific study combined with the experience of years.

You have done well if you noticed three, but look again because there are six in total.

A better technique would be to use a highlighter pen and perhaps you would have noticed all six.

Scanning

As a child, and in truth as an adult, I always looked forward to receiving the Radio Times magazine a few weeks before Christmas. In those days this was the only means of finding out which television programmes were scheduled over Christmas. However, I never actually read it but simply scanned through it, stopping whenever I noticed something of interest or something that took my fancy, like a favourite film. This is what is meant by scanning.

Intensive reading

Intensive reading involves focusing on all aspects of a written passage, especially the comprehension or meaning behind what has been written. Oral examinations are assessed in this way; after reading a specific passage students are asked explicit questions concerning the events and characters involved. Very much the who, what, when and where.

Engineering procedures are also written in a logical way, in what is known as a step-by-step approach. However, I have personally witnessed the failure of a piece of electronic equipment, what is referred to as a black box, and, convinced that it was faulty, I actually replaced it. When the replacement component also failed I reviewed the procedure, realising that I had missed out several fundamental steps regarding key switches and levers. All this would have been indicated on the test procedure and even highlighted as Pre-Test Requirements. But human nature being what it is, certain impulses tend to kick in, such as being over eager to finish the task. What happens is that you actually spend more time trying to fix a fault which has only occurred because of a failure to read the testing procedure correctly or intensively in the first place.

Extensive reading

This is similar to intensive but is more about reading for pleasure. It is possible to combine some of these techniques; for instance, by skimming a text to get an overview before reading intensively to obtain more detail.

Top tip ⭐

Intensive reading focuses on all aspects of written text. If you are reading an engineering procedure or exam question – read everything carefully, twice if you have to.

Learning outcome 5.1

Outline the different ways in which people learn.

It is believed that people learn in a variety of ways and this idea stems from a theory sometimes referred to as 'learning styles'. The belief is that there are three categories of learners and are identified through the acronym VAK:

V: is for visual

A: is for auditory

K: is for kinaesthetic

Visual learners learn through images or anything they can observe such as diagrams, pictures, films, even images drawn on flip-charts. There is also a belief that some dyslexic people think in pictures.

Auditory learners process learning through the sound of the spoken word, which means they must combine hearing and speaking when processing information.

Kinaesthetic learners process information by actually feeling and handling objects; in other words, their best way of learning is actually engaging in 'doing' something.

Caution should be used here. Learners do not necessarily sit neatly in these categories and will use all three elements, but they may favour one or even prefer two of the learning styles over the other. Equally a person might use all three in equal measure. The object of the exercise is to enable the learner to establish what kind of emphasis works best for them, and that could be a single learning method or a combination.

Top tip ⭐

Learning styles.

V: is for visual

A: is for auditory

K: is for kinaesthetic

I refer to learning styles when a student is experiencing difficulty in understanding certain principles, when I will aim to re-explain but using their preferred method of learning. Equally, if you can establish your own preference – perhaps as a visual learner, for instance – you can catalogue a whole host of effective and appropriate visual images. Furthermore, uncovering some appropriate YouTube videos, which by their very nature include visual, sound and a small element of kinaesthetic (doing), would also be a very useful learning technique.

Learning outcome 5.2

Identify some barriers to effective listening.

Learning outcome 5.3

Contribute ideas and ask questions.

Effective listening is when an individual fully understands what is being said to them, as opposed to someone who simply nods their head pretending to understand. Listening is definitely an acquired skill because it involves focusing and maintaining your auditory attention. It might be helpful to think of it as active listening which means paying attention not only to **what** someone is saying, but also to **how** they are saying it and especially if they are emphasising or repeating any particular aspect.

Body language is very important when communicating, especially maintaining good eye contact. You should also try to speak clearly but never in an aggressive manner, otherwise the other person is less likely to be receptive of your tone of voice. This is a perfect example of how your own behaviour can discourage or anger another person. Equally damaging is stopping someone else from speaking or expressing their opinion; it is a sure way of preventing any meaningful conversation. Contrast this with an active and effective listener who patiently and politely allows a person to speak and then comes back with meaningful and positive suggestions. This is why, for communication to be effective, it is vital that all parties involved feel valued – they need to feel that their contribution and opinion matters.

Some people are naturally shy while others love the sound of their own voice. But they can both invest in one very important characteristic: respect. This means being respectful of others' opinions even if they differ drastically from your own. Equally, when asked for your opinion or when you are given an opportunity to share your ideas, grasp it. It signifies that you are willing to contribute and are not looking to hide behind other people.

I find the following analogy very useful when likening two different styles of communication by comparing a helicopter and seagull.

A helicopter is a person who rises above a discussion and thinks about what the discussion is about and how they can add something of value to its thread. A person who acts like a helicopter is hovering overhead so that the 'wood can be seen from the trees'.

A seagull, on the other hand, is 'someone who flies in, circles around, makes a lot of noise, swoops down and dumps on a few people before quickly flying off again'.

Saying

Can't see the wood for the trees: when you are unable to understand a situation clearly because you are too involved in it.

Top tip

Try to be respectful of other people's opinions, even if they differ from your own.

Top tip

Think about your tone of voice. Sometimes offence is taken not by what is said, but how it's said.

It is important to share your viewpoint, but consider the what, why, who and how:

- What are you going to say
- Why are you saying it (to offer your opinion)
- Who am I speaking to (formally to a client or informal to a friend)
- How am I going to say it (verbally, formal letter, email, etc.)

The practice of asking questions should relate to a job interview. Not only should an applicant expect to answer certain questions in relation to the vacancy they have applied for, but they are normally given an opportunity to ask a question of their own at the end of the interview. This is where a well-thought-out question can leave a positive and lasting impression on the interviewer.

Learning outcome 5.4

Outline the possible barriers to contributing to a group activity or discussion.

Some people will feel comfortable conducting and taking part in a group discussion. This could be because they are naturally outspoken, love to perform or a type of person that thrives in hearing the sound of their own voice. This can create a barrier to active or effective group activity, since less enthusiastic learners might feel intimidated and reluctant to join in. Equally, some learners, even when they are given an opportunity, prefer to hang back and be led by the discussion, but this means that they, too, are missing out on an opportunity to contribute.

A balance has to be struck, in that everyone involved must be encouraged to engage in the conversation, even if they do not agree with everything that has, or is, being said. It is vital that different opinions are expressed. This is how a court of law works – different opinions and questions are posed, which will hopefully allow the truth to emerge. Even a jury is permitted to ask questions, largely through the judge, if they are unsure of anything in relation to a point of law. But more importantly, when the jury meets at the end of the trial to discuss the evidence presented, it is vital that all the jurors, even the naturally quiet ones, express their opinion.

Learning outcome 6.1

Identify suitable methods of retaining information.

I stated previously that a lot of short periods of study are generally more effective than one very long period. This is because you will start off refreshed after a mini-break but also you are reviewing the material more frequently which helps your brain to function effectively.

We all have two types of brain function, called short term and long term memory. We only learn if we can transfer information from our short term memory into our long term memory and the best way of doing that is to create meaning. If meaning can therefore be attached to an element of study, then there is more chance of a person remembering it. We will therefore discuss a number of different ways of creating meaning, but like most techniques, you need to find the one that works best for you.

Top tip ★

A lot of shorter study periods tend to be better than one large session because after a mini-break your concentration levels start off on a high.

Ask a friend or family member if they have ever driven somewhere but then on arrival or at a point during the journey they suddenly realised that they have no memory of how they got there. If they answer yes, and many of them will, tell them not to panic – their brains are acting normally. Our short term memory retains about seven unrelated facts at any one time, but dumps anything that is not memorable. Therefore, if you are driving on the same road day by day and nothing out of the ordinary happens, these particular memories are discarded.

Speak out loud

This technique basically draws on two learning styles, visual and auditory, and it recommends that reading out loud can create greater stress and emphasis, which hopefully will be more memorable. Some people use this kind of technique because it helps them remember if they turned the cooker off or locked the front door. I have also seen a dyslexic individual who, when leaving for home, will start to speak to themselves: 'wallet, keys, phone'. They have learnt this technique because over the years they have left one of these items behind.

Mind maps

Mind maps are effective as both a learning tool and study technique because when trying to learn a subject, it gives a learner a sense of the whole concept (idea) as well as breaking it down into individual elements. Teachers can create mind maps to help their students – for instance, at the beginning of Chapter 5 of this book there is a description of why an electrician is a unique trade given all the different systems and material they have to work with. A mind map can be

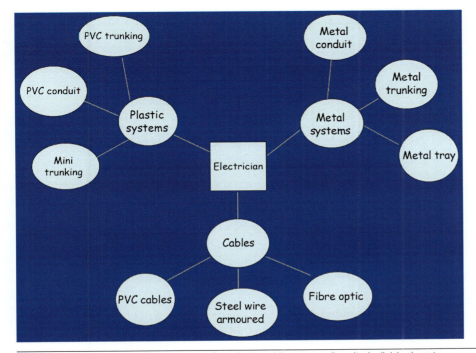

Figure 6.3 Mind maps offer a sense of a whole subject as well as its individual parts

used to signpost the same information as can be seen in Figure 6.3. Mind maps can also be an extremely useful studying strategy especially if student create and design it themselves. Why not give it a go? Think of a subject and start by writing the name of the subject in the middle and then attaching various appropriate branches.

Mnemonics

A mnemonic is a technique used to remember facts by using the first letter in a sequence of words and creating a rhyme.

If you were taught Pythagoras' Theorem in school then you may well recall SOHCAHTOA, which is a clever way of remembering all the trigonometry relationships with right-angled triangles. SOHCAHTOA is an example of the use of a mnemonic, but it is far better if the student can create their own. This is because a personal mnemonic creates individual meaning.

I once taught a 52-year-old taxi driver who had enrolled on to a full-time Level 2 Electrical Installation course. Unfortunately, SOHCAHTOA meant nothing to him, so I encouraged him to set about creating his own mnemonic. The order of the letters involved with SOHCAHTOA can be split into three as long as the following letters are kept together:

CAH SOH TOA

What follows is a personal mnemonic to him that reflected what he did for a living and the fact that he had a smallholding – some land with a few animals. Looking over the cluster of letters he came up with his own version:

Cabbies are happy (CAH)
Sitting on horses (SOH)
Till others arrive (TOA)

This was a personal mnemonic, which in all probability will not work for anyone else, but that does not matter. What I find interesting is that the person in question can still recall the mnemonic six years after completing the course. This is because a personal mnemonic is able to uncover an intimate connection by drawing on memories or vivid images, which in turn creates meaning.

Mnemonics can also be used to learn new difficult words including how they should be spelt. 'Extraneous' is a mouthful for anybody to learn and especially to spell, but applying a little rhyme might help. Firstly, break the word into two parts:

Extra and neous

To remember the end of the word we can create a little rhyme:

EXTRA + No (N) Elephants (E) or (O) Unicorns (U) Seen (S)

Flashcards

Using flashcards is a learning technique which involves creating pairs of cards, either by using blank playing cards or cut-out paper versions. Use one card to write an important term, and on another card write a definition or an important fact about that term. An example is shown in Figure 6.4. Create several to cover a particular subject, then turn them all over and mix them up. Flip over two cards. If they match you get to keep them; if not, you have to turn them over again. Keep going until you match them all.

Top tip

Mnemonics are especially effective if students can come up with their own versions.

Top tip

Creating meaning also creates memory and learning.

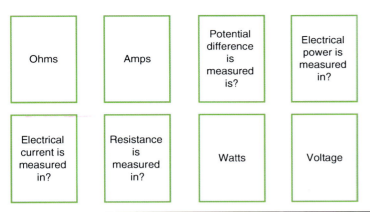

Figure 6.4

Alternatively, you can write 'start' on one side of a card (or piece of paper) and then write out a question on the back of the same card. The next card should have the answer to this first question and a new question. This continues until you have matching number of questions and answers as well as the word 'finish' as shown in Figure 6.5. Once created, mix them all up, and look for the card with the word 'start' written on it. Flip it over and look at the question on the other side. Try to find the matching card with the answer and continue until you navigate to the finish.

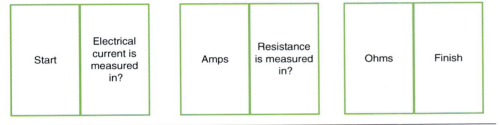

Figure 6.5

Hot Potatoes

Hot Potatoes is a free resource that teachers can use to create learning activities for their students. However, students can also use this resource as a learning aid.

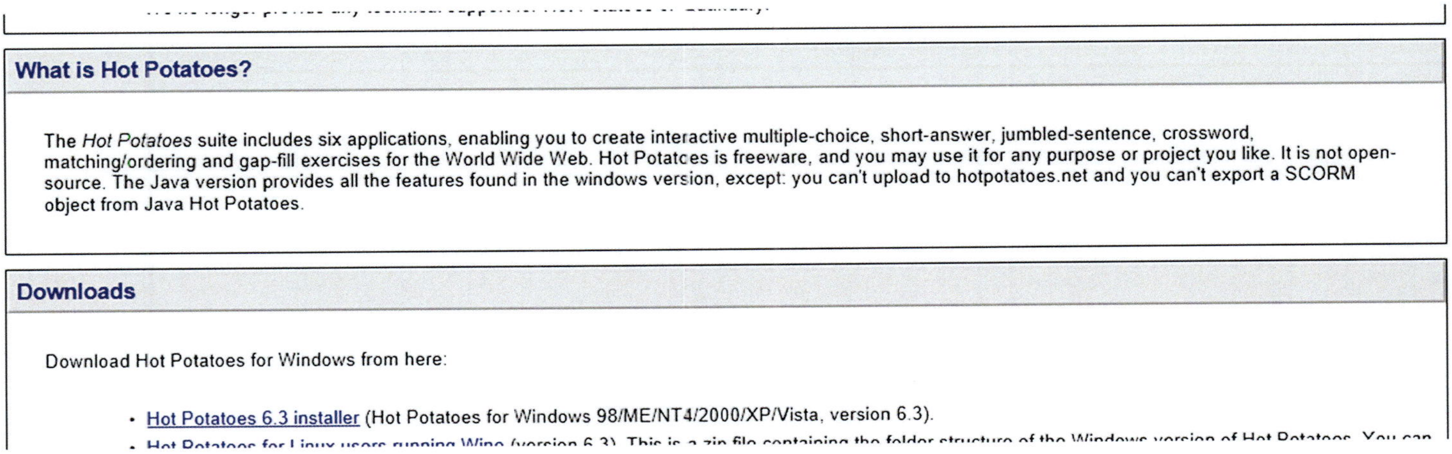

What is Hot Potatoes?

The *Hot Potatoes* suite includes six applications, enabling you to create interactive multiple-choice, short-answer, jumbled-sentence, crossword, matching/ordering and gap-fill exercises for the World Wide Web. Hot Potatoes is freeware, and you may use it for any purpose or project you like. It is not open-source. The Java version provides all the features found in the windows version, except: you can't upload to hotpotatoes.net and you can't export a SCORM object from Java Hot Potatoes.

Downloads

Download Hot Potatoes for Windows from here:

- Hot Potatoes 6.3 installer (Hot Potatoes for Windows 98/ME/NT4/2000/XP/Vista, version 6.3).
- Hot Potatoes for Linux users running Wine (version 6.3). This is a zip file containing the folder structure of the Windows version of Hot Potatoes. You can

Figure 6.6

Figure 6.7

To download the programme, click on to the Hot Potatoes home page: http://hotpot.uvic.ca/

This will bring the following page up and select **Downloads** as indicated in Figure 6.6.

You then need to select the specific software or operating system for your computer such as: Hot Potatoes for Windows 98/ME/NT4/2000/XP/Vista, version 6.3.

Once this is done, follow the set up procedure (I would recommend that you create a desktop icon).

Figure 6.8

The Hot Potatoes menu should open as shown in Figure 6.8; if not, click on the desktop shortcut.

Then select the following application: **JMatch**

A file will open such as that shown in Figure 6.9.

Enter a list of key terms in the left hand side (Left (ordered items)). Then enter the matching answers down the right hand side (Right (jumbled items)).

Select **File**, then use **Save as facility** to store the programme.

This will be a master version of the programme, which will allow you to add or delete questions in the future. The master programme is green in colour.

Once you have added all your information, to use the activity you will need to create and save it as a webpage.

Select **File**, **create web page**, then choose **Drag/Drop Format** (Ctrl + F6)

The **Save As** window will open, so save application in a location of your choice.

As you save, the programme will ask you if you want to 'View the exercise in your browser'.

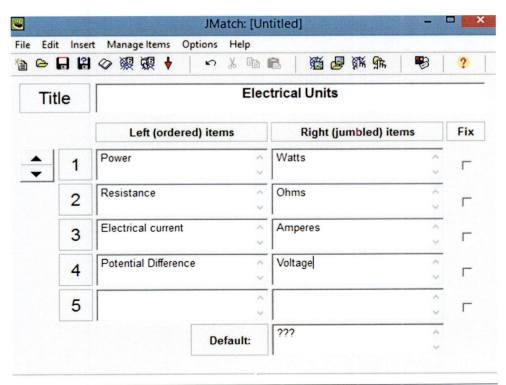

Figure 6.9

Alternatively, you can view it by double clicking on the explorer file.

Figure 6.10 shows what the **Drag and Drop** facility looks like when opened as an Explorer file.

Note:

1. You might need to select: **Allow blocked content**
2. Continue with loading programmes and select **Yes** if you encounter pop-ups about titles containing spaces. Normally this does not cause a problem

The added bonus of creating your own activities is that EAL use this kind of assessment as part of their online examinations.

Collaborative learning

There is an old saying, which never fails to annoy me:

Those who can, do; those who can't, teach.

This is mostly stated by people who have never taught and know nothing of the difficulties involved in teaching, especially when the great majority of educators try to make a difference to the lives of students. However, I can honestly state that it was only after I started to teach that I fully began to understand certain elements of electrical principles. This is because students ask all kinds of really interesting and sometimes strange questions and occasionally I have to research the answer.

Collaborative is a big word which simply means working alongside someone else – a study buddy. Why not challenge a fellow student or friend that is on the same course as you? Arm yourself with the same textbook and have a game of penalty shoot-out.

Look though a particular chapter and each of you write down five **fair** questions. Only include questions that you can answer yourself. It's also a good idea to

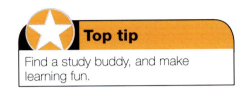

Top tip

Find a study buddy, and make learning fun.

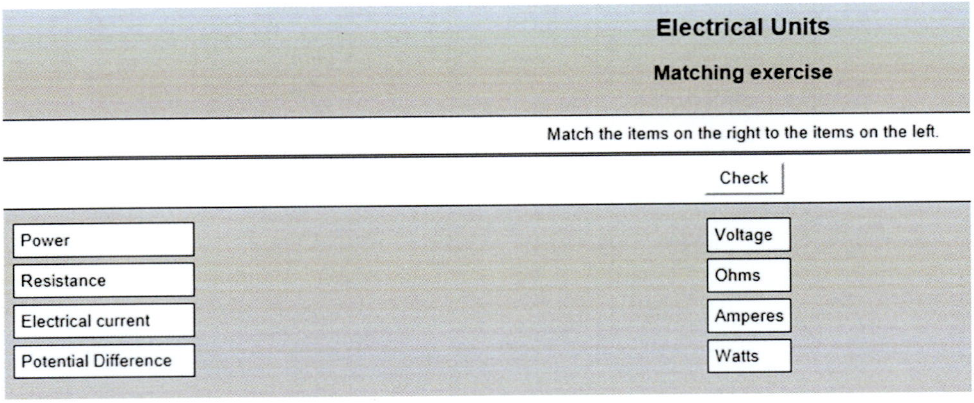

	Electrical Units
	Matching exercise

Match the items on the right to the items on the left.

Check

Power		Voltage
Resistance		Ohms
Electrical current		Amperes
Potential Difference		Watts

Figure 6.10

write down the page number to prove where the answer lies in the book. Take it in turns to ask a question.

A correct answer scores a goal. Best out of five wins!

Why not give it a go?

Look-Cover-Check

This technique is very good when you want to review and learn certain images, text or terms from a glossary. Look at each in turn, cover them up, and then try to recall them. Keep going until you can remember all the terms or come to a point when you cannot learn anymore.

Chunking numbers

It is known that learning is easier if we follow patterns, and this applies to numbers.

Try to remember the following number sequence:

795367451206782

Most people would struggle, but they might find that chunking the numbers up will be considerably easier.

795 367 451 206 782

In other words, if you have to learn a sequence of numbers, break the numbers into chunks of three.

Top tip

Most people chunk up their new mobile phone number when they are trying to learn it.

Learning outcome 6.2

Demonstrate effective strategies for taking notes.

Note taking should not be just a method of scribbling down information on a piece of paper. It should be done in such a way so that, when you look over it, the information should still make sense.

There are many effective strategies for taking notes, but, first things first, always store them in a logical order within a file and do not just tip them into the bottom of your bag or deposit them into a locker. You can also use coloured separators to fit between each chapter, which makes it easier for you to concentrate on certain elements of study.

Top tip

As long as you gain permission, some students record their lectures.

Some people find it helpful to write down the actual date of the lesson/lecture and even the room or location of the lecture. This again might help them remember and jog their memory when they are revising, especially if the teacher or lecturer has used or presented very useful information or powerful analogies that can be revisited by the student.

Mind mapping was covered previously as a study technique, but is equally as effective in summarising information or as a way of planning and structuring essays.

Another useful technique is to ask your tutor for a copy of their PowerPoint presentation but in a three-slide view format as shown in Figure 6.11. This is a very effective way of taking notes, because it allows you to write down a key point or word against each slide, rather than engaging in a lot of writing.

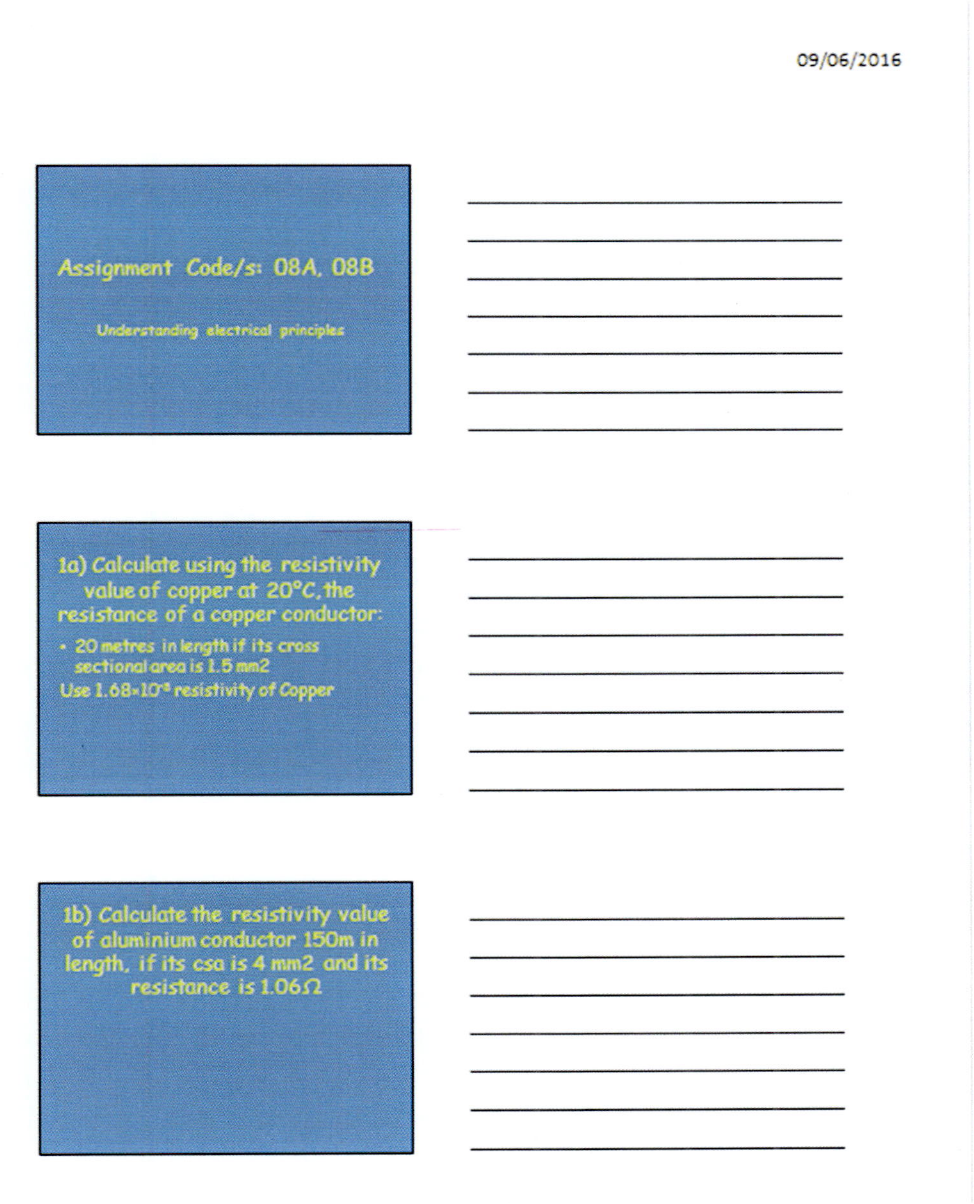

09/06/2016

Assignment Code/s: 08A, 08B

Understanding electrical principles

1a) Calculate using the resistivity value of copper at 20°C, the resistance of a copper conductor:
- 20 metres in length if its cross sectional area is 1.5 mm2
Use 1.68×10^{-8} resistivity of Copper

1b) Calculate the resistivity value of aluminium conductor 150m in length, if its csa is 4 mm2 and its resistance is $1.06\,\Omega$

Figure 6.11 PowerPoint in 3 slides per view allows a student to highlight some key points rather than having to engage in a lot of writing

Another great tip involves using a highlighter pen to emphasise any key words or phrases on tutor-issued handouts.

Packing in a lot of studying and embracing a range of studying techniques will only get you to a certain point, but in reality if you are unable to absorb the information in the first place, you are topping up very little information. Dyslexia, for instance, is not related to intelligence and some of the difficulties that dyslexics experience extend to far more than difficulties with reading. For instance, a lot of the issues will be related to limitations with short term memory and finding difficulty in paying attention. Other students might experience personal problems or are holding down several jobs as well as trying to study and learn.

If you are experiencing difficulties, it is very important that you discuss such issues with your tutor in order to see what can be done. On the flip side, if you find a particular activity very helpful or enjoyable, then also discuss this with your tutor so that they are made aware of the usefulness of this particular technique or activity.

I personally provide a summary to all my students before each and every exam. Although there are no guarantees that all the students will read or in some cases even look at them, I do find they are helpful to certain individuals who are feeling anxious about an upcoming examination. If this sounds like you, there is no harm in you asking for the same from your tutor.

Another useful method to support learners is a learning journal, which is a book given to students by the tutor. It allows the student to make notes in private, away from other students, to highlight what worked well in that particular college week, but, more importantly, what they found difficulty with. This relies more on the student than the teacher, because the teacher cannot respond if the book remains empty.

Last, many colleges provide additional study opportunities, which are specifically put on to coincide with end of year examinations. These sessions will also cover techniques that are particularly useful when studying.

Multiple choice exams

Many awarding bodies assess through online examinations, with the majority, but not all, of the questions being asked in multiple choice format. In the past, these sorts of questions were designed in a certain well-known format. Normally, a question would contain four possible options, but in truth only two of the options were likely, such as that shown below:

A: The answer

B: Close to the answer but not likely

C: Unlikely

D: Highly unlikely

These days all the options have to be possible or at least linked in some way to the answer. This also applies to calculations since they will include options that have basically inverted the correct formula. Consider the following question about Ohm's law.

Question: If 5 amps of current flow through a 2 Ω resistor, calculate the supply voltage.

The correct formula to calculate the supply voltage is V = I x R. This should work out as:

V = 5 x 2 = 10 volts.

Top tip

A learning journal is very effective for those students who do not like to ask questions in the classroom.

Top tip

Always read exam questions very carefully and look out for any key words.

But please note that one of the options would be:

Voltage = $\frac{2}{5}$ = 0.4 volts (formula inverted).

When sitting an online exam, students are given an examination slip, which contains their personal details and an individual code that they enter to start their assessment. Because they are permitted to write on this paper, I always tell my students to use the back of this slip in order to draw certain useful objects such as the Ohm's law triangle. This can help the student to extract the correct formula and reduce the chance of inverting it. This is especially important because, for some reason, some students always assume that the highest value in the question is always inserted at the top of the formula.

Other techniques in relation to online examinations include:

1) Ensure you read the question very carefully, students tend to skim or scan the question, but what is required is intensive reading.

2) Look for a key word in the actual question, which will indicate which of the options is more favourable. This is shown from the example below:

   Which of the following personal protection equipment should be worn in very dusty environments?

   a. face mask

   b. respirator

   c. silk handkerchief

   d. face screen

   Perhaps the silk handkerchief is a weak option, but a face mask and screen are possible answers. The key words in the question are 'very dusty' and respirators include a filter which will screen out any dust to stop the person inhaling it.

3) A far as question writing goes, the use of the word 'Not' is not recommended. This is because it is easily missed. Although the word **NOT** should be shown in capital letters and bolded out as shown here, if the reader misses it then they have very little hope of making a successful choice. Therefore remember to apply point number 1 above and read the question very carefully.

4) A valuable technique or sequence to use is to look at the question, then look at all the options in turn, then go back to the question, before you select. Repeat this sequence if required.

5) Having followed the previous stage, if you are unsure of the answer, most online examinations include a flag option, which allows a student to opt to return to the question.

6) Sometimes a question writer will use a long or difficult word and this can be very off-putting for a student. However, quite often, if the question is read again but this time omitting the big word, in most cases, the question will still make sense.

7) Ask your tutor for some sample papers sometimes called mock exams. You might even find some online or within your Moodle facility.

8) Ensure that you eat a good breakfast on the day of the exam. It is also vitally important that you keep yourself hydrated by sipping water, even during the exam if this is allowed. This might mean that you will need a toilet break, so bear this in mind before you enter the exam room.

9) Last, if you notice some students leaving early, don't panic. Go by your own pace. In my experience those students who leave early gain the lowest marks.

Summary

A lot of the techniques discussed and offered in this chapter may be wildly different to your usual methods of study, but they can offer alternative strategies, especially if your usual methods are not working. What is very important is that you adopt a positive mindset, no learning programme will be effective unless an individual commits to a study programme. This starts by writing down your exam dates and thinking seriously about what techniques you are going to use and when and where are you going to start to use them.

Education can empower you to greater success, but your teachers can only do so much; it is largely up to you to put the effort in.

I was a serious underachiever in school and did not gain any GCSEs, or O Levels as they were called then, and I even managed to achieve a U grade for two of my subjects. U stands for Unclassified; in other words, if I had been playing darts, I failed to hit the dart board. I now realise that I have dyslexic tendencies; any diagnosis in my case made that more difficult because I am an excellent reader and, in fact, can speed read very well. My difficulties stem from my working memory (I have to read something several times before it sinks in), which is why I find it difficult to master tasks straight away – I always need multiple attempts. I also find taking verbal instructions extremely difficult, especially since my mind wanders easily – I now realise that I also suffer from episodes of attention deficit.

And yet, I am a qualified lecturer and an ex-RAF Chief Technician, currently involved in adapting certain technical books and the actual author of this particular version. But I still suffer doubts about my abilities, especially when I am struggling to succeed at something new – such as learning a new skill. When this happens I am propelled back to when I was at school, re-experience a lot of negative thoughts, and feel trapped and distressed. However, I never gave up. I actively looked for ways and techniques that could help me to remember. I dismissed any negative comments I received and I tried that little bit harder or longer at learning something. Remember, the letter 't' is the only difference between the words 'can' and 'can't'. 'T' stands for try!

I firmly believe that my past failures make me a better teacher today. Therefore, I offer you a challenge: find your own techniques, battle on and think about this famous saying – 'the harder you work, the luckier you become'.

Unit: QMSA1/01

Chapter 6 checklist

Learning outcome	Assessment criteria – the learner can:	Page number
1. Understand the demands of a course of study.	1.1 Identify the main aim of a course of study. 1.2 Outline the demands of a course of study including: Timescale Attendance Assessment criteria Forms of assessment Self study	230 230
2. Understand how to organise study time effectively.	2.1 Identify appropriate times of study. 2.2 List the key features of a good, safe and productive learning environment. 2.3 Identify the personal challenges that may affect study.	230 231 231
3. Understand how to prioritise and set realistic targets for study.	3.1 Outline what SMART targets are. 3.2 Identify SMART targets for own study.	232 232
4. Be able to find and use information relevant to the course of study.	4.1 Use a range of reference systems to locate specific information. 4.2 Use a range of reading techniques for different purposes.	233 235
5. Understand how to listen in and contribute actively to a learning environment.	5.1 Outline the different ways in which people learn. 5.2 Identify some barriers to effective listening. 5.3 Contribute ideas and ask questions. 5.4 Outline the possible barriers to contributing to a group activity or discussion.	236 237 237 238
6. Keep information in a usable format.	6.1 Identify suitable methods of retaining information. 6.2 Demonstrate effective strategies for taking notes.	238 244

Preparing for an interview

Learning outcomes

The learner will:

1. Know information required to prepare for an interview.
2. Be able to prepare for interview questions.
3. Be able to plan travel for an interview.

EAL Electrical Installation Work – Level 1. 978-1-138-23206-8

Learning outcome 1.1

Identify the purpose of the interview.

Learning outcome 1.2

Outline the key information about the job/placement/course drawing on application information.

The purpose of an interview is to satisfy the aims of two people: an employer who has an opening or vacancy, and prospective employees who are looking for employment.

The actual interview should be a two-way conversation, in which both parties have their individual aims.

Employer perspective

As far as the employer is concerned, they are trying to decide if the person in front of them is what they are looking for and will endeavour to check out certain qualities to find out if that's the case. They will also look at what qualifications, skills and potential the candidates offer, as well as personal attributes such as:

- Positive attitude
- Strength of character
- Previous experience

The employer wants to know if an applicant can do the job or position being advertised and the applicant needs to convince them that they and not the other people being interviewed is the right person to fill the vacancy.

The interviewee also needs to find some answers to some important and key questions such as:

- Do I want this job?
- Can I do this job?
- Does this job offer me the opportunities I want?

Curriculum vitae (CV)

It is incredibly important that you adapt your CV for any position you are applying for, trying to identify similar responsibilities you might have held in other jobs or even positions in clubs and organisations. However, although your CV and cover letter will normally have a major influence on your chances of being invited to interview, if they contain wildly exaggerated claims or an entry such as 'my interests include wrestling with my brother' then your interview will be short, sweet and unsuccessful. Remember that a prospective employer will look through your CV, and will question you on the information before them.

For those not yet fortunate to receive an interview and perhaps have yet to put together a CV and cover letter, examples have been provided in Appendix D at the end of this book. Also included is a basic guide to grammar that covers some of the primary rules, which will hopefully ensure that your application will be as accurate as possible.

Important point

An employer carries out an interview to see if any of the candidates are the right person to fill a vacancy.

Top tip

A cover letter is very important, because it allows you to personalise your application.

The interview

There could be more than one person conducting the interview, with each person likely to ask different questions. For instance, a technical person might want to identify and verify your engineering experience, interests or qualifications such as maths and English. On the other hand, a member of the Human Resources (HR) department is more likely to ask questions such as:

- How much notice do you have to give your current employer?
- Are you aware of the salary we are offering?
- How long have you been at your current position?
- Why do you want to leave?

It very much strengthens your chances of being given serious consideration if you know the answers to these questions, but your answers should never be flippant. For instance, stating that you hate your current job and hate your employer will not get you noticed in any meaningful way. Contrast this to an applicant that states: my current job involves a lot of travelling or the hours are long. Alternatively, my job is not challenging or that it has no real possibility of promotion.

Learning outcome 2.1

Prepare answers to questions that might be asked at the interview.

There are hundreds of different questions that could be asked but some are more common than others. Typical questions that employers ask include:

- Why have you applied for this job?
- What do you know about the vacancy?
- How did you hear about this vacancy?
- What can you offer our company?
- What do you know about this company?
- What are your strengths?
- What are your weaknesses?
- What are your long term goals?
- Where do you see yourself in five years?

Top tip

If you were asked 'what can you offer?', how would you respond?

The way you answer these types of questions could well decide if you are either given the position or, in some cases, shortlisted for a further interview. Remember, you are effectively a brand, and therefore you need to sell yourself.

It is very important that you carry out research in order to identify what the job involves but also, if possible, find out relevant information about the company. For instance, were they known by a previous name, have they been taken over by a larger organisation? This is very important because some companies are:

- Regional (involved in one specific area)
- National (only based in one country)
- Multinational (based in various countries and even continents)

Top tip

Carry out some research about the company that you want to work for!

Carrying out some research prior to your interview and arming yourself with this kind of knowledge will impress and strike a positive chord with the interviewer.

Top tip

If you enjoy stripping down computers or are constantly using your spare time to work on bikes and cars, inform the interviewer – it indicates that you are mechanically minded.

Before the interview, you should also make a list of everything that you have achieved or accomplished in your life. Again, even as a school leaver you will be surprised what some young adults have not disclosed at the interview or put down on their CV for that matter, which would have impressed me enormously. One young person went swimming before school, getting up at 5 a.m., swimming a mile or two, and then attending school. And yet he failed to mention this impressive show of commitment. The employer needs to gauge who you are as a person, which, to be fair, is a very difficult task in such a short amount of time. However, disclosing, for instance, that your hobbies include working on bikes, cars and such items – if true, of course – would indicate that you are a very active, hands-on and mechanically minded person.

I always ask the following question to all prospective electrical installation candidates:

- Tell me something about yourself that will make me sit up and take notice and make me want to give you a place on the course.

I fully appreciate that it is a difficult question, especially if you are 16 years old. But what I am looking for are examples such as:

Top tip

Before your interview write out a list of all relevant accomplishments and experiences.

- Raising money for charities
- Winning a competition of any sort
- Holding down a part-time job whilst at school or college
- Working over weekends or holiday periods
- Helping out a family member who is elderly or disabled
- Being a member of the Scouts, Beavers, St John's Ambulance Brigade
- Being a member of a sport's club or team
- Being a member of a choir
- Representing your school, county or country at a sport
- Martial arts or boxing achievements
- Duke of Edinburgh Award
- Helping out on a farm
- Being a member or helping out Search and Rescue or Coastguard organisations
- Working as a volunteer in any shape way or form

What many of these activities represent is commitment and dedication but also experience of working with people, dealing with customers, taking orders, being trusted to handle money and accepting responsibility. Other attributes that an employer will look for include an applicant that is well-versed with Health and Safety procedures or a person who willingly accepts discipline and is self-motivated. An example that always sticks in my mind was a young man who said that he had taken on a second part-time job because his father had fallen ill and could not work.

Motivation is very important and most prospective employers will look favourably on a self-motivated individual who is willing to go that extra yard – a person that can be trusted to not only get themselves up and report to work on time, but be given certain responsibilities. The old saying that a team is only as strong as its weakest link is very true. The interview is about uncovering if you are the weakest link or a potential asset. But from an applicant's perspective, it can also be a way of letting the interviewer know that you are very definitely a cooperative person and very much a team player.

Top tip

A lot of employers look for attitude before aptitude.

The interviewer will also sometimes ask different types of questions, whereby an applicant is given a situation and then certain questions will be asked to see how they respond. For instance:

- If a customer was being extremely difficult and aggressive, how would you respond?
- What action would you take if you witnessed an electrician being electrocuted?
- What would you do if came across two employees arguing, one of whom was being abusive?

There is no one single answer to these types of questions, but common sense can be used. For example, even though a customer might be difficult, there is an old saying: the customer is always right. Therefore, arguing and shouting back, even with the most trying of people, is not an option.

Instead, calmly telling them that you will refer the matter to a more senior person is a far more acceptable course of action. Most companies who deal with customer care will have a no violence policy. They are common in hospitals, unemployment offices, shopping outlets and pubs, and knowing about such things can arm you to respond to the question by stating calmly to the customer, 'Please lower your voice and remain calm, I am doing my best to help you!' **The interviewer is basically trying to find out about your people skills. How good a communicator are you? What kind of procedures and policies are you aware of?**

Figure 7.1 Prospective employers will be very impressed if an applicant has some knowledge of basic emergency procedures

In the case of the second question, most people have been at one time inside a laboratory or workshop – either when in school, college or through previous employment – where red emergency switches (Figure 7.1) are installed. Equally, in most situations, most people would know to use a wooden implement such as a brush in order to remove any live supply away from the person. As long as, of course, you do not put yourself in danger. **The interviewer is trying to find out if you are aware of basic Health and Safety procedures as well as trying to see if you can think under pressure.**

How to deal with a tense argument between two people is never an easy question to answer or actually observe in real life. However, if the argument is becoming abusive, then you should report such matters to a superior. Sometimes, the people involved are not aware that they are actually shouting, therefore in some circumstances trying to talk to the people involved in a calm manner could actually defuse the situation. **The interviewer is trying to find out how you feel about bullying and if you are strong enough as a person to speak up against such matters.**

If you are ever asked a question and you are unsure of the answer, rather than saying nothing, it is far better to reply: I am not sure of the correct answer but I would go and find my supervisor to seek further advice.

Why not carry out a mock interview with yourself or a friend and prepare and practice answering similar questions?

Learning outcome 2.2

Identify questions to ask which show interest in the job, placement or course.

Top tip

Asking a relevant question at the end of an interview will give you credibility as a candidate.

Anyone attending an interview should expect to not only answer questions but also ask questions themselves. Depending on what details were disclosed in the job advert, the following are typical of what you could ask:

- What is the hourly pay?
- What is my holiday entitlement?
- What are my prospects of promotion?
- When can I expect to hear from you?
- Is there a probation period?
- Does the job come with any benefits?

Asking these types of questions indicates to any prospective employer that you have thought seriously about this position – especially if you are a school leaver – because it helps them view you as a young adult trying to make the transition to the world of work. Job benefits could include a company pension, which is in some cases non-contributable, meaning the employee does not have to contribute. Other companies offer private health care schemes, reduced sport club membership fees and many more such benefits. Even if you are attending an interview for a part-time or zero hours contract, the more that you know about a company, or at least have the presence of mind to ask, the greater your chances of making a positive impression.

Learning outcome 3.1

Confirm the time and place where the interview will be held.

Learning outcome 3.2

Plan a route and means of transport to arrive on time for the interview.

I once arrived early for my first day at work, or so I thought, before realising that I had gone to the wrong place. The college had two locations in the same town and a similar sounding name, but it could easily have been avoided – if I had planned the journey properly. I could have blamed my sister-in-law; after all, she was the person who had given me imprecise directions. But, at the end of the day, it was my job – and my responsibility!

I had failed to plan properly and although I had not planned to fail that's exactly what happened. Had I been late for an interview, however, then I would have been told in no uncertain terms to go away and never darken their door again.

With the range of technology that exists today – which includes 24-hour news, weather and travel, satellite navigation and AA Route Planner (a free Internet resource, through which you can print off a route) – there should be little scope for error. There is, however, always the possibility of roadworks and unexpected diversions – sheep on the road if you live in Wales as I do – or, heaven forbid, a road traffic accident. Public transport such as buses and trains can be delayed, so it is important not only to plan your route but add some extra time for the journey, even if it means you arrive early – you can always go for a cup of tea or a toilet break.

That said, I did all of these things but I still ended up arriving late, simply because I had not read the letter properly and relied completely on the opinion of another person. It's always a good idea to bring your letter of invitation to interview along on the day. Security is very important these days, and the letter can clearly state why you are entering a building. This is especially important if it involves Ministry of Defence land. However, an interview letter will also include the location's full address, so should you get lost you can show it to a passer-by in order to get some directions. Because of the unknown factor many people carry out a mock journey prior to their interview. This way they can see how long or difficult the journey will be, as well as investigate car parking options.

Top tip

Plan your route to the interview very carefully and allow extra time for disruption.

Certain businesses, especially national and multinational organisations, will actually carry out telephone interviews first and only successful candidates will then progress to the next round. If you are given a time for such a phone interview, then I think it goes without saying that you want to ensure that the line is free to receive the call. You also want to prepare in the same way as has already been discussed, so that you are armed with certain information and you are fully aware of what position you have applied for and what you have to offer. Certainly, a lot of companies will expect you to tell them either your projected grades or GCSE results already achieved.

Top tip

Take all relevant qualifications and achievements with you to the interview.

Often, before or after an interview, many organisations will give the applicants an aptitude test. There are many different kinds, but in my experience they will definitely include testing an applicant's literacy and numeracy skills. There are many free Internet resources available so that you can brush up on your skills, but none better than BBC Skillswise. The website includes a lot of mathematics activities but some are delivered through interactive games, which means you can not only entertain yourself but learn at the same time. Most aptitude tests are geared around questions that are not necessarily difficult, but instead are designed to see how many you can complete over a short period of time. Therefore, before the interview, practice some of these activities especially those involving basic fractions, long multiplication, long division and percentages. It will soon come back to you.

It is also important to note that preparing for an interview also means you should dress appropriately. When you dress smartly it gives the interviewer the impression that you have made an effort. This does not mean that all applicants should necessary go out and buy a suit, although this does help, but as a

Top tip

Give yourself every chance. Dress for success.

minimum, wearing clean trousers, a shirt and perhaps even a tie indicates that you are taking the interview seriously. It also makes you stick out for all the right reasons, secretly making the interviewer take notice as soon as you walk in.

In other words: dress for success!

I once interviewed a young lady with shocking pink hair for an engineering position. Thankfully, I always try to look at the person and disregard their fashion preferences. She was very articulate, a single mother who wanted to better herself and also help the life prospects of her child. Her drive and determined attitude won me over. She is now a supervisor at the same company, but I have often wondered how many interviewers would have dismissed her as an applicant on the basis of the colour of her hair, despite perhaps having been a punk, mod or rocker themselves back in the day (if these terms mean nothing to you – ask your parents or grandparents).

The task of an interviewer, however, is not easy, especially since you are only given a short period of time to make a judgement call. During my time as an interviewer, two other people stand out for me, both for the wrong reasons. The first was a young man who decided to keep his MP3 player connected in his ear piece throughout the interview and only took it out to ask me to repeat a question. The other decided to chew bubble gum all the way through the interview, and what made matters worse was that he chewed with his mouth open! Neither did themselves any favours. Chewing gum could well have been because he was nervous and, let's face it, most people are. But try and find other ways of relaxing, such as picturing the interviewer on the toilet.

Eye contact

It is incredibly important that you maintain eye contact with the interviewer; you need to be believable and you need to be yourself! To bolster your chances further, there are some very useful objects or items that you can take with you. For instance, school reports, attendance records and certificates of achievements, especially for those involving the Duke of Edinburgh Award or First Aid, are very impressive. A handy tip is to put all your certificates in a posh

Figure 7.2 Put all your qualifications in a posh looking folder and take it along to the interview

Figure 7.3 If you have fabricated something such as a piece of engineering, take it along to the interview

looking folder (Figure 7.2); it creates an impression that you are a very organised person who is also very proud of their achievements.

I was also mightily impressed with a candidate who brought along a piece of engineering metal work that they had personally fabricated (Figure 7.3) – well worthwhile if the job or position you are expressing an interest in involves technical work. Such tactics cannot fail to impress those on the other side of the desk.

Summary

If you receive a letter of invitation to attend an interview, then you have already been successful, since many fail even to get that far. However, do not waste the opportunity; carry out proper research on the job and the company.

Practice answering some interview type questions as well as brushing up on some basic mathematics. Take your qualifications with you on the day. Plan your route and decide how you are going to get there, how long the journey takes and where you are going to park.

Dress for success!

I was never a Boy Scout but I do think their motto is right: 'Be prepared'.

Being ready for a job is one thing, but ask yourself: Are you job ready?

Chapter 7 checklist

Learning outcome	Assessment criteria – the learner can:	Page number
1. Know information required to prepare for an interview.	1.1 Identify the purpose of the interview. 1.2 Outline the key information about the job/placement/course drawing on application information.	252 252
2. Be able to prepare for interview questions.	2.1 Prepare answers to questions that might be asked at the interview. 2.2 Identify questions to ask which show interest in the job, placement or course.	253 256
3. Be able to plan travel for an interview.	3.1 Confirm the time and place where the interview will be held. 3.2 Plan a route and means of transport to arrive on time for the interview.	256 256

Synoptic assessment

The following sets of questions are a combination of Health and Safety, environmental protection measures, electrical installation theory and electrical principles.

1. Which of the following is **NOT** considered a renewable source of energy?
   a. wind
   b. rain
   c. tides
   d. natural gas

2. The Health and Safety Act 1974:
   a. applies to employers
   b. only affects employees
   c. applies to clients only
   d. only affects visitors

3. How much voltage is dropped across R2?

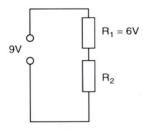

   a. 15 V
   b. 3 V
   c. 6 V
   d. 9 V

4. Which of the following would be considered essential when working at height?
   a. safety harness
   b. safety boots
   c. safety gloves
   d. safety high visibility clothing

5. What does SWL mean?
   a. safety working level
   b. safest working level
   c. safe working load
   d. safe working limit

6. Which of the following is the best conducting material?
   a. copper
   b. aluminium
   c. gold
   d. silver

7. Good housekeeping is important in the workplace because:
   a. it improves the accident rate
   b. it stops accidents
   c. it increase the number of accidents
   d. it reduces the number of accidents

8. What colour label is associated with a CO_2 fire extinguisher?
   a. red
   b. blue
   c. cream
   d. black

9. Which one of the following is used to determine the electrical current in a circuit?
   a. amperes x watts
   b. volts x ohms
   c. watts x amperes
   d. $\dfrac{\text{volts}}{\text{ohms}}$

10. A ladder should be installed at an angle of:
    a. 57°
    b. 100°
    c. 75°
    d. 25°

11. Which of the prefixes below is represented by kilo?
    a. 100
    b. 1000
    c. 1000,000
    d. 10

12. Electric current is caused by movement of:
    a. atoms
    b. neutrons
    c. electrons
    d. protons

13. The **MAIN** purpose of a cable sheath is to:
    a. improve its conduction properties
    b. improve its insulation properties
    c. provide protection from mechanical damage
    d. improve the cable's flexibility

14. An employer must carry out a risk assessment to identify:
    a. potential hazards
    b. potential pollution
    c. potential controls
    d. potential problems

15. The purpose of a protective device in a circuit is to:
    a. control the circuit
    b. operate to isolate the circuit

c. indicate when the circuit is dead

d. indicate when a fault occurs

16. Which of the following is **NOT** a statutory document?

a. Electricity at Work Regulations 1989

b. Electricity Safety, Quality and Continuity Regulations 2002

c. IEE Wiring Regulations and On Site Guide

d. Health and Safety at Work Act 1974

17. Two resistors both 20 Ω in value are connected in series. Calculate their total resistance:

a. 20 Ω

b. 10 Ω

c. 40 Ω

d. 0 Ω

18. The SI unit for resistance is the:

a. watt

b. ohm

c. volt

d. amp

19. A battery produces:

a. a mains supply

b. direct current

c. alternating current

d. pulsed a.c.

20. If an electrical circuit is supplied with 6 volts and draws a current of 10 A, calculate the circuit power:

a. 60 A

b. 60 W

c. 0.6 A

d. 0.6 W

21. Which of the following is **NOT** a fossil fuel?

a. oil

b. gas

c. coal

d. nuclear

22. If an electrical circuit has a resistance of 5 Ω and draws a current of 10 A, calculate the circuit voltage:

a. 2 V

b. 50 V

c. 0.5 V

d. 15 V

23. Air pollution can produce:

a. acid air

b. acid rain

c. alkaline rain

d. alkaline air

24. Statutory regulations are acts of law, therefore they must:
   a. be ignored
   b. be given serious consideration
   c. be obeyed
   d. be reported

25. Which of the following is negatively charged and causes electricity?
   a. protons
   b. electrons
   c. neutrons
   d. atoms

26. Earthing protects a circuit by operating a:
   a. circuit switch
   b. circuit insulator
   c. circuit protective conductor
   d. circuit breaker

27. Landfill should be used:
   a. as a first resort
   b. as a last resort
   c. occasionally
   d. frequently

28. Which of the following is **NOT** a major component in an electrical installation circuit?
   a. emergency battery
   b. fuse
   c. light bulb
   d. wall switch

29. Which of the following environmental systems extracts solar heat that has been retained in the ground?
   a. air source heat pump
   b. micro-hydro
   c. ground source heat pump
   d. micro-turbo

30. Where a long horizontal run of conduit is to be marked out, which of the following would be most appropriate?
   a. metal ruler
   b. spirit level
   c. plumb line
   d. chalk line

31. What is the most common modern insulator for domestic twin and earth cable?
   a. Polyputthekettleon
   b. PolyVinylChloride
   c. PolyVinylpaint
   d. PolyVinylCyanide

32. Every time we buy goods, travel by car, bus, train or aeroplane
    we generate:
    a. carbon dioxide
    b. oxygen
    c. hydrogen
    d. nitrogen

33. Why do modern domestic properties include a separate ring for the
    kitchen?
    a. modern kitchens do not contain a lot of electrical equipment
    b. modern kitchens contain a lot of electrical equipment
    c. modern kitchens include electric cookers
    d. modern kitchens contain electric radiators

34. Which **TWO** of the following are associated with electrical
    power circuits?
    a. radial
    b. ring
    c. one-way lighting
    d. two-way lighting

35. Which of the following would be considered as hazardous waste?
    a. cardboard
    b. fluorescent tubes
    c. plastic conduit
    d. metal conduit

36. Which of the following results would indicate a good electrical
    connection?
    a. $0.01\ \Omega$
    b. $240\ \Omega$
    c. $1\ M\Omega$
    d. $100\ \Omega$

37. The type of cable that is normally installed inside domestic premises
    would be
    a. SWA
    b. Twin and earth
    c. MICC
    d. PVC single

38. Which of the following environmental systems is classified as
    zero carbon?
    a. air source heat pumps
    b. wind power
    c. solar PV
    d. CHP

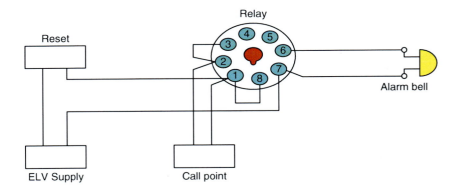

39. What kind of diagram is shown here?

    a.   circuit diagram

    b.   wiring diagram

    c.   block diagram

    d.   layout diagram

40. Grey water recycling reuses which **TWO** of the following:

    a.   rainwater

    b.   toilet water

    c.   sink water

    d.   bath water

Answers to synoptic assessment				
1. D	2. A	3. B	4. A	5. C
6. D	7. D	8. D	9. D	10. C
11. B	12. C	13. C	14. A	15. B
16. C	17. C	18. B	19. B	20. B
21. D	22. B	23. B	24. C	25. B
26. D	27. B	28. A	29. C	30. D
31. B	32. A	33. B	34. A, B	35. B
36. A	37. B	38. B	39. B	40. C, D

Answers to 'Test Your Knowledge'

Chapter 1

1. D	2. A	3. A	4. B	5. A
6. C	7. C	8. D	9. A	10. C
11. A	12. B	13. B	14. B	15. C
16. C	17. B	18. C	19. B, D	20. D
21. D	22. B	23. C	24. D	25. D
26. D	27. D	28. D	29. A	30. D
31. A	32. D	33. B	34. B	35. B
36. C	37. A, D	38. C	39. B	40. D
41. C	42. A	43. C	44. D	45. C, D

Chapter 2

1. A	2. D	3. C	4. A	5. B
6. D	7. D	8. D	9. B	10. A
11. D	12. C	13. D	14. C	15. A
16. A	17. B	18. D	19. A	20. D
21. B	22. A	23. D	24. D	25. B
26. C	27. B	28. B	29. A	30. A
31. B	32. A	33. D	34. A	35. B, C
36. D	37. A	38. A, D	39. B, D	40. C
41. B	42. B	43. D	44. C	45. A
46. C	47. C	48. C	49. B	50. C
51. B	52. B	53. A	54. D	

Chapter 3

1. A, C	2. B, C	3. A, C	4. C	5. B
6. D	7. A, C, D	8. D	9. D	10. A
11. B, C	12. D	13. A, C, D	14. C	15. B, C
16. A, D	17. A, D	18. A, C	19. A, B, C	

Chapter 4				
1. C	2. D	3. C	4. B	5. C
6. B	7. D	8. D	9. B	10. A
11. D	12. C	13. B	14. C	15. D
16. B	17. B	18. D	19. A	20. C
21. C	22. A	23. A	24. A	25. C
26. D	27. C	28. A	29. D	30. C, D
31. B	32. A	33. C, D	34. A, C	35. A
36. B	37. C	38. D	39. A	40. C
41. D	42. C	43. B	44. B	45. A
46. D				

Answers to chapter activities

Chapter 1

Requirements of Health and Safety

Question 1: The Health and Safety Act 1974 is **NOT** a legal document. (False: it is a legal document)

Question 2: Certain work-based regulations are known as the Six Pack Regulations. (True)

Question 3: Employees should, by law, have to provide their own PPE. (False: employers must provide PPE in relation to hazards that are present in a particular environment)

General health and safety activity

The Health and Safety Act 1974 = 1	Manual handling = 2	Enabling Act that acts as a health and safety umbrella = 1	Act that ensures employers train their employees with correct lifting techniques = 2
Personal Protection Equipment = 3	Workplace Health and Safety Act = 4	Act that ensures employers provide PPE = 3	Act that ensures employers provide a safe working environment = 4

Safety sign activity

Mandatory sign = 1	Prohibition sign = 2	Blue 'must do' = 1	Red and white 'must not' = 2
Warning sign = 3	Safe condition sign = 4	Yellow and black 'be aware of hazard' = 3	Green 'safety information' = 4

Health and Safety systems

Question 1: Wearing gloves or using barrier cream can reduce dermatitis. (True)

Question 2: A mandatory sign should be thought of as 'must do'. (True)

Question 3: COSHH is a system that controls the use of substances. (True)

Personal protection equipment activity

Hard hat = 1	Hand protection = 2	Head protection = 1	Gloves = 2
Ear defenders = 3	Safety glasses = 4	Steel toecaps = 5	Eye protection = 4
Safety shoes = 5	Other than gloves this can protect against dermatitis = 6	Barrier cream = 6	Hearing protection = 3
Hi-viz vest = 7	Protect your whole body = 8	Be seen, be safe = 7	Overalls = 8

First aid and safety procedures

Question 1: Your first action when attending to a casualty is to ensure you are not in danger. (True)

Question 2: It is better to remove a hazard than wear PPE. (True)

Question 3: Burns caused by heat or chemicals should be treated by applying cold running water for at least 10 minutes. (True)

Question 4: As an employee you must be aware of designated escape routes as well as nominated assembly points. (True: they should be explained through an induction process)

Question 5: Asbestos is safe as long as you do not get it wet. (False: moving it or disturbing asbestos especially with activities such as drilling causes tiny fibres to be released. Spraying asbestos material with water will reduce the amount of fibres released but it is still not safe to work on)

Safety systems

Question 1: Oxygen, fuel and heat form the fire triangle and you need to remove all three to stop a fire. (False: you only need to remove one)

Question 2: Anyone can put up scaffolding. (False: only qualified and registered scaffolders are permitted to put up scaffold)

Question 3: Overhead cables are dangerous because they have a very thin layer of insulation. (False: overhead cables tend not to be insulated other than through air)

Question 4: All hand-held equipment used on a construction site must be powered through a 110 V supply. (True)

Question 5: Only serious accidents have to be recorded. (False: all accidents have to be recorded on accident forms, but serious accidents have to be further recorded through RIDDOR)

Five of the Six Pack Regulations

Management of Health and Safety at Work Regulations 1999: A regulation that ensures employers provide safe systems of work A = 2

Provision and Use of Work Equipment Regulations 1998: A regulation that ensures an employer provides proper work-based equipment B = 1

Manual Handling Operations Regulations 1992: A regulation that ensures employers train people how to lift loads safely C = 5

Workplace (Health, Safety and Welfare) Regulations 1992: A regulation that ensures that all workplaces are safe D = 3

Personal Protective Equipment at Work Regulations 1992: A regulation that ensures that when hazards cannot be removed PPE is provided E = 4

Workplace hazards and good working practices

Main cause of accidents in the construction industry: Slips, trips and falls A = 3

Main cause of deaths in the construction industry: Working at height B = 4

Good husbandry practices: Daily clean up and ensuring that tools and equipment are maintained correctly C = 5

Asbestos: Dangerous material that not made any more but is still fitted to some buildings. Dangerous if disturbed D = 2

Barrier cream and PPE can prevent this skin condition: Dermatitis E = 6
You notice that some oil has spilt on the floor. What should you do?: Raise the alarm by warning others and stand guard near the hazard F = 1

Workplace procedures

Use a sack trolley when carrying heavy loads A = 5

Near miss: Even an incident that did not result in harm must be reported in order to learn from the event B = 1

Recovery position: Put an unconscious casualty in this first aid posture C = 3

Assembly point: Report here in an emergency D = 4

CO_2 extinguisher: Type of fire extinguisher that is OK to use with electrical fires E = 2

Chapter 2

SI units activity A

Ampere = 1	Voltage = 2	Electrical current = 1	Potential difference = 2
Power = 4	Energy = 3	Joule = 3	Watt = 4
Length = 5	Time = 6	Resistance = 7	Degrees Kelvin = 8
Ohms = 7	Temperature = 8	Metre = 5	Second = 6
Coulomb = 9	Mass = 10	Electrical charge = 9	Kg = 10

SI units activity B

Put an X in the correct box	SI base unit	SI derived unit	Electrical unit
Time	X		
Electric current (amperes)	X		X
Coulomb		X	X
Length	X		
Area		X	
Voltage		X	X
Power		X	X
Mole	X		
Mass	X		
Temperature	X		
Resistance		X	X

Multiples, sub-multiples and prefixes

milli = 1	micro = 2	1×10^{-3} 1/1000 = 1	1×10^{-6} 1/1000000 = 2
nano = 3	pico = 4	1×10^{-9} 1/1000000000 = 3	1×10^{-12} 1/1000000000000 = 4
kilo = 5	mega = 6	1×10^{3} 1000 = 5	1×10^{6} 1000000 = 6
terra = 8	giga = 7	1×10^{12} 1000000000000 = 8	1×10^{9} 10000000000 = 7

SI units

Question 1: The SI unit for energy is the watt. (False: Watt is the SI unit for power. The Joule is recognised as the SI unit for energy)

Question 2: milli is a submultiple and can be thought of as 1 ÷1000 (1/1000). (True: 1 divided into a thousand bits)

Question 3: kilo is a prefix which represents 1000 or 1×10^{3}. (True: as in 3 kV, three thousand volts)

Ohm's law

Question 1: When a circuit becomes live, current flows through a component and voltage appears across it? (True)

Question 2: Resistance helps electric current. (False: resistance opposes electrical current)

Question 3: This is the Ohm's law triangle. (True)

Electrical supplies

Question 1: A domestic mains supply is associated with direct current (d.c.). (False: the mains is an alternating current supply)

Question 2: A transformer only works when supplied with an a.c. supply. (True)

Question 3: We can extract the main formula for power for an electrician by using this triangle. (True)

Conductor/insulator activity

Put an X in the correct box	Conductor	Insulator
Copper	X	
Air		X
Steel	X	
Iron	X	
Ceramic		X
Paper		X
Thermoplastic material		X
Gold	X	
Wood		X
Carbon	X	
Magnesium Oxide		X
Aluminium	X	
Thermosetting material		X

Conductors and insulators

Question 1: The movement of electrons is what gives us electricity. (True)

Question 2: Gold is the best conducting material. (False: silver is the best conducting material)

Question 3: Air is used as an insulator. (True)

Fill in the gaps

In a SERIES circuit the CURRENT is common to all components. To work out the total resistance you simply ADD the resistance values. Christmas tree lights used to be wired in series, but one major disadvantage is that if one light bulb BLOWS, all the lights go OUT.

In a PARALLEL circuit the VOLTAGE is common to all components. A very important rule for parallel circuits is that the total resistance is always LESS than the smallest value. Modern Christmas tree lights tend to be wired in parallel, because if one light bulb blows, the others remain ON.

Electrical installation knowledge

Question 1: Voltmeters should be connected in parallel. (True)

Question 2: A wiring diagram tells you how a circuit operates. (False: a circuit diagram tells you how a circuit operates, a wiring diagram tells you how a circuit should be wired)

Question 3: Loop-in and junction box are two methods of wiring domestic lighting. (True)

Question 4: Any number of sockets can be fitted within a ring circuit. (True: the only limitation is that the room cannot be any bigger than 100 m² in area)

Question 5: Modern domestic properties should be installed with two ring circuits. (False: it should be three: downstairs sockets, upstairs sockets and kitchen area)

Electrical SI and derived units

Ampere: Electrical current A = 1

Coulomb: Electrical charge B = 4

Voltage: Potential difference C = 6

Ohm: Electrical resistance D = 3

Watt: Electrical power E = 2

Joule: Energy F = 5

Electrical terms and items

Water: Can be compared to an electrical current A = 2

Direct current: Type of electricity used in batteries B = 4

Mains voltage: Domestic 230 V single phase supply C = 3

Ohm's law triangle D = 5

An electrical cable can be compared to: A water pipe E = 6

Transformer: Item used to step voltage up and down. Only works with alternating current F = 1

Chapter 3

Carbon classification activity

	High carbon	Low carbon	Zero carbon	Generates electricity by producing steam
Solar photovoltaic		X		
Natural gas	X			X
Hydroelectric			X	
Coal	X			X
Solar thermal		X		
Biomass		X		
Wind			X	

Environmental technology activity

Solar thermal	Solar PV	System that uses the sun to produce heat	System that uses the sun to produce electricity
= 1	= 2	= 1	= 2

Biomass = 4	Hydrogen fuel cell = 3	Fuel is made by combining hydrogen and oxygen = 3	Material derived from living, or recently living organisms. = 4
Air Source heat pump = 5	Ground source heat pump = 6	Uses pipes installed in the ground = 6	Opposite to a fridge, extracts heat from the air outside = 5
CHP = 7	CCHP = 8	Fuel in: heat and electricity out = 7	System that combines elements of cooling, heat and power = 8

Environmental technology

Question 1: Wind turbines do not contribute to global warming because they do not generate any carbon dioxide. (True)

Question 2: Solar PV can be thought of as a transducer because it changes one form of energy into another. (True)

Question 3: A hydroelectric system (running water) is renewable. (True)

Reuse or recycle activity

Reusable left-over lengths of conduit can be = 1	Plastic bottles = 2	can be recycled = 2	reused = 1
Landfill is a = 4	Hazardous waste = 3	Fluorescent tubes and batteries are classified as = 3	Last resort = 4
Rain collected to water your garden and washing your car = 5	Grey water recycling = 6	Reusing tap water and recycling it to flush toilets = 6	Rainwater recycling = 5

How we can change the way we live

Question 1: Thermoplastic material can be recycled. (True: thermosetting materials, however, cannot)

Question 2: Recycling helps the environment because fewer items have to be made. (True: because every time we produce materials and goods we increase our carbon footprint)

Question 3: If everyone makes small change in their life this can help create a big meaningful impact on the environment. (True)

Microgeneration technologies

A system that installs pipes in the ground and extracts heat from the soil: Ground source heat pump A = 2

A system that changes solar energy to create hot water: Solar Thermal B = 4

Fuel in, heat and electricity out: Combined heat and power C = 5

System that changes solar energy to electricity: Solar PV D = 3

A system that uses free flowing water to create electricity: Micro-hydro E = 1

Let's help the environment

Landfill: Where materials that cannot be reused or recycled are sent to A = 3

Every time we buy goods, travel by car, bus, train or aeroplane we generate: Carbon dioxide B = 1

Fluorescent tubes and batteries: Types of equipment classified as hazardous waste C = 2

Built environment: Material that we use to build houses and other buildings D = 5

The total amount of greenhouse gases produced by a person's activities including travelling and the food that they consume is known as their: Carbon footprint E = 4

Chapter 4

Tools activity

Tape measure = 1	Pencil = 2	Item used for marking out wooden products = 2	Used to measure something out = 1
Hammer = 3	Screwdriver = 4	Used to tag cables in place = 3	Used to tighten terminations = 4
Wire strippers = 5	Spirit level = 6	Used to check for level = 6	Used to strip insulation = 5
HSS = 7	Spade drill bit = 8	Drill bit used on metal = 7	Drill bit used to drill holes through wooden joists = 8

Fixings activity

Plastic plug = 2	Rawlbolt = 1	Fixing used with brick = 2	Fixing used with masonry = 1
Spring toggle = 3	Gravity toggle = 4	Fixing used with plasterboard = 3	Fixing used with long vertical walls = 4

Tooling

Question 1: A chalk line is a type of marking out device that is appropriate when measuring long horizontal surfaces. (True)

Question 2: A masonry drill bit would be appropriate choice when drilling holes through a wooden joist. (False: to drill holes through wood a spade drill bit should be used. A masonry bit should be used when drilling through stonework)

Question 3: A Rawlbolt type fixing is appropriate when securing equipment on hollow walls. (False: a Rawlbolt is a type of fixing used on stonework, a spring toggle should be used on hollo walls)

Protective device activity

Protective device that operates through heat (thermal operation) = 1	Protective device that operates with either heat or magnetism = 2	Circuit breaker = 2	Fuse = 1
Additional protection device that only operates through magnetism = 3	Type of circuit breaker used in domestic environments = 4	RCD = 3	Type B = 4
Type of circuit breaker used with industrial equipment that generate high in-rush currents = 5	Type of circuit breaker used in commercial fluorescent lighting = 6	Type D = 5	Type C = 6

Electrical installation terms and knowledge

Question 1: PVC flat profile cables are secured in place using PVC clips. (True)

Question 2: Recessed lights are known as downlighters and are normally fitted with either tungsten halogen or LED lamps. (True)

Question 3: A metal gas pipe is known as an exposed conductive part. (False: any metal object that does not form part of an electrical installation is known as an extraneous conductive part)

Question 4: Earthing provides a low resistance path for a fault current to flow, which will cause a protective device to operate very quickly and isolate the circuit. (True)

Wiring systems activity

PVC flat profile = 1	PVC/SWA = 2	Domestic wiring = 1	Sub-main = 2
Flexible Conduit = 3	Tray/ladder = 4	Used with motors to offset vibration = 3	Metal systems used to support SWA cables = 4
Plastic conduit = 5	Metal conduit = 6	Type of enclosure used in commercial environments = 5	Type of enclosure used in industrial environments = 6

Industrial and commercial environments

Question 1: Plastic conduit is far stronger than metal conduit. (False)

Question 2: Industrial environments often route electrical cables overhead and away from the floor area. (True)

Question 3: Conduit is wired through non-sheathed PVC singles. (True)

Testing activity

Earthing = 1	Exposed conductive part = 2	Circuit protective conductor = 1	Metal switch box = 2
Healthy value for continuity of c.p.c. = 3	Minimum acceptable insulation resistance value = 4	0.05 Ω = 3	1 MΩ = 4

Documentation

Question 1: BS 7671 the Wiring Regulations are legal documents. (False: but they are written in such a way that they comply with all statutory requirements. The Wiring Regulations are seen as a code of practice)

Question 2: If you are building a new house, the Building Regulations can be ignored. (False)

Question 3: By always referring to BS 7671 an electrician is complying with their code of practice. (True)

Roles in the construction industry activity

Architect = 1	Clerk of work = 2	The eyes and ears of the architect on site = 2	Person who creates drawings and plans = 1
Joiner = 3	Electrician = 4	Craftsperson who builds things by joining pieces of wood = 3	Craftsperson who installs electrical systems = 4
Plumber = 5	Gas fitter = 6	Craftsperson who installs gas fires, cookers and boilers = 6	Craftsperson who installs drinking water, sewage and drainage systems = 5
Plasterer = 8	Ground worker = 7	Craftsperson who is employed to prepare a site for the shallow foundation of a new home = 7	Craftsperson who forms a layer of plaster on an interior wall or ceilings = 8
Tiler = 10	Decorator = 9	Craftperson who tiles out kitchens and bathrooms = 10	Craftsperson who applies paint and wall paper = 9

Communication

Question 1: An architect produces a set of plans and technical drawings. (True)

Question 2: When deliveries arrive on site, they should always be checked against a purchase order. (True: always check your original purchase order against your delivery note. Then physically ensure that you have been given the correct quantity and type of materials)

Question 3: Most accidents involve a breakdown of communication. (True)

Question 4: Data protection means that you are allowed to give away personal information with consent. (False: never disclose someone's personal information)

Wiring systems and enclosures

Thermoplastic PVC sheathed and insulated cable (flat profile): Cable type used in domestic properties A = 6

Flexible conduit: This type of wiring enclosure is used as the final connection to a motor circuit B = 3

Mineral insulated (pyro): Cable type used in high temperature and explosive environments C= 2

PVC/SWA: Cable type with very high level of mechanical protection and is often installed underground as a sub-main D = 1

Bus-bar system: Power system used in industrial environments that require high current use E = 5

Plastic conduit: This type of wiring enclosure would need to be warmed up to room temperature before being used if left outside overnight F = 4

Drills and fixings

HSS drill bit: Type of drill bit that should be used with metal A = 3
Masonry drill bit: Type of drill bit that should be used with stonework B = 4
Spade drill bit: Type of drill bit that creates holes in wood C = 5
Spring toggle: Type of fixing used in hollow walls such as plasterboard D = 1
Rawlbolt: Heavy duty type of fixing used to secure items in stone E = 2
Plastic plug: Type of fixing used with brick F = 6

Components

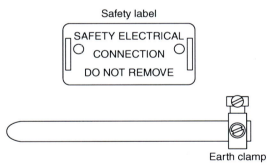

Clamp – British Standard BS951 A = 4

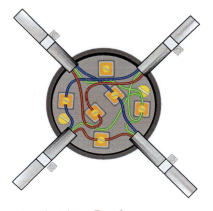

Junction box B = 3

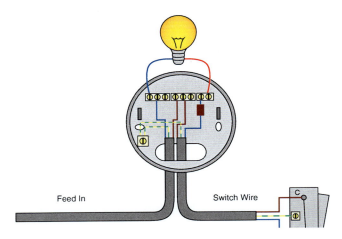

Ceiling rose – loop-in wiring system C = 1

Earthing sleeving required for twin and earth cable D = 2

Chapter 5

Occupation task activity

Electrician = 1	Plumber = 2	Installing cables = 1	Installing pipes = 2
Work with wood products = 4	Work with plaster = 3	Painter and decorator = 7	Matches building to the requirements of different businesses = 8
Architect = 5	Quantity surveyor = 6	Plasterer = 3	Joiner = 4
Work with paints, varnish and wall paper = 7	Facilities manager = 8	Produces detailed plans and drawings = 5	Works out material costs = 6

APPENDIX
Formulas

Ohm's law

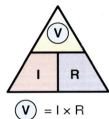 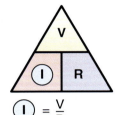 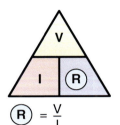

$V = I \times R$ $I = \dfrac{V}{R}$ $R = \dfrac{V}{I}$

Electrical power

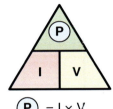 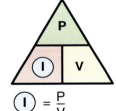 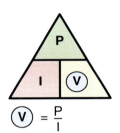

$P = I \times V$ $I = \dfrac{P}{V}$ $V = \dfrac{P}{I}$

Electrical principles formulae transposed

1. Ohm's law $V = I \times R$

    • Transpose for $I = \dfrac{V}{R}$

    • Transpose for $R = \dfrac{V}{I}$

2. Power = Voltage (V) x I (current)

    • Transpose for $I = \dfrac{P}{V}$

    • Transpose for $V = \dfrac{P}{I}$

3. Resistors in series $Rt = R1 + R2 + R3$

4. Resistors in parallel $\dfrac{1}{Rt} = \dfrac{1}{R1} + \dfrac{1}{R2} + \dfrac{1}{R3}$

APPENDIX
Useful websites

Health and Safety Executive: http://www.hse.gov.uk/

UK government website: https://www.gov.uk/

BBC Skillswise English (Literacy): http://www.bbc.co.uk/skillswise/english

BBC Skillswise maths (Numeracy): http://www.bbc.co.uk/skillswise/maths

Calculate how big your environmental footprint is: http://footprint.wwf.org.uk/

The Construction Index: http://www.theconstructionindex.co.uk/

APPENDIX
Credibility of information

It is often said that we live in a knowledge society where we create, share and use a huge amount of information to improve our lives. Take the internet, for instance – it houses a wealth of information but, equally, we are bombarded by spam email, nasty viruses, internet scams and even when we do safely and successfully navigate on to a web address it might contain pop-ups.

In this age of technology, the amount of information we are exposed to can often be overwhelming. How can we possibly tell what's 'credible'?

Question: When a website ends in '.com' – like bt.com – what does this mean?

Were you aware that '.com' actually means .company; in other words, this website is trying to sell you something? Therefore, just because you have gained access to a website, this does not necessarily mean that you should believe everything you see and act on it without question. Even when you are simply carrying out research, try to make a judgement about whether the information before you is credible!

This also applies to newspapers, who often create sensationalised headlines which most people, but not all, can see through. For instance, believe it or not, certain tabloids have run the following:

- MUM GIVES BIRTH TO 8 POUND TROUT
- TALKING BRUSSEL SPROUT
- LORD LUCAN SPOTTED ON MISSING SHERGAR

Other headlines, written by so-called distinguished newspapers, also called broadsheets, have in the past conveyed secret and hidden meanings. Headlines such as:

ANOTHER CANNABIS TEENAGER IN KNIFE KILLING

will, in the minds of certain people, imply that all teenagers are knife-wielding maniacs.

Equally damning is:

BLACK MEN TO BLAME FOR MOST VIOLENT CITY CRIME

Although this article did state that black men are twice as likely to be victims of such crimes, it failed to state that black men are more likely to have been born into poverty, which in turn can lead to a disjointed education and no real career opportunities. In other words, the article decided to use the effects of certain social problems but ignore the root cause. Problems, however, occur because when people read these types of headlines they often believe them without question. In other words, if a headline or well-known label makes sense, then it must be true.

Equally, such labels are sometimes given to young adults, because in the minds of certain people, young adults lack experience, therefore they must be empty vessels. You can see how this plays out with school leavers. At 16, and with their parents' consent, they can legally marry and have a child. If they earn enough they have to pay tax. Again, with their parents' consent they can join the armed forces and be issued with a deadly weapon. However, they will not be eligible to vote until a further two years have elapsed.

Credible research

Having now established that certain information might include biased opinions and hidden messages, it goes without saying that before deciding if a book, article, website, or other resource you locate is a valid and credible resource ask yourself six questions:

Who?

Who is the author? What are his/her credentials?

What?

What information can you use from this resource? Is it valid?

Where?

Where did the author(s) get the information? Is it their personal viewpoint?

When?

When was the resource produced? Was it published recently? (A lot of websites include a 'created on' or 'last updated on' date.)

Why?

Why do you think this resource exists? For instance, is the resource trying to sell you something? Is it biased? Is it trying to change your opinion?

How?

How comprehensive is the resource? Does it go into the depth you need?

Validity is also another key word, and involves asking has anybody approved this information. For instance, online forums and Wikipedia are very useful information sites, but it's difficult to establish the credentials of those who have posted the information.

Certain sites are credible because they contain policies, announcements, publications, statistics and consultations involving various agencies, public bodies and government run departments. For example check out: www.gov.uk/

Another valid and credible online resource can be found at the Health and Safety Executive (HSE) website: www.hse.gov.uk/

The HSE website search facility can be used to research a variety of procedures such as risk assessments and method statements; all very useful just before an interview. To view what kind of interesting information that is freely available enter the following descriptions into a search engine and check out the links:

Electrical injuries: www.hse.gov.uk/electricity/injuries.htm

Company sentenced after employee received electric shock: http://press.hse.gov.uk/2015/company-sentenced-after-employee-received-electric-shock/

Electrical safety: www.hse.gov.uk/toolbox/electrical.htm

Leeds firm in court after trainee's electric shock: http://press.hse.gov.uk/2014/leeds-firm-in-court-after-trainees-electric-shock/

The HSE web-site also includes a very useful section called Myth of the Month. This has been designed specifically by the HSE to counter how the construction of information can lead to misinformation.

According to the HSE, over the years, many an urban myth or what is known today as fake news has been passed into the public domain, claiming that Health

Top tip

Remember, as previously recommended, use quotation marks to narrow your search:
'Electrical installation jobs'.

and Safety Inspectors have banned or outlawed a variety of activities, all for the greater good.

Why not check out the Top 10 worst health and safety myths by visiting: www.hse.gov.uk/myth/top10myths.htm

APPENDIX
Basic grammar

The importance of using correct grammar and spelling when writing and compiling a CV and cover letter cannot be overemphasised. Any prospective employer will be less than impressed with an applicant who sends in an application littered with errors.

What follows, therefore, is a basic guide to grammar that covers some of the primary rules, but will, hopefully, ensure that you give yourself every chance of being selected for interview and not fall at the first hurdle.

Sentences

A proper sentence should contain a **subject** (the sentence must be about someone or something).

The subject is usually a noun and there are four different types:

Proper noun – for example, names of people, organisation, places and things (proper nouns are capitalised)

Common noun – a broad general term, such as car, city or teacher (common nouns are not capitalised)

Collective noun – for example, gaggle of geese or herd of cattle

Abstract noun – for example, hate, happiness or love (in other words, feelings)

A proper sentence should also contain a **verb** (a doing word or action).

For example, the following is a proper sentence: **Pete has wired the house.**

> Whom is the sentence about? **Pete**.

> What is he doing? **Wiring the house**.

Instead of Pete you can also use a pronoun – for example, I, me, he, she, you, etc.

Spell checkers do sometimes spot an incorrect sentence as well as incorrect spellings. Look out for the subject/verb agreement error message.

Sentences should start with a capital letter and end with a full stop. They are really basic errors to make, but are made nonetheless!

Commas can be used to separate words and word groups in a simple series of three or more. For example:

> **My personal experience includes: customer care, adopting health and safety procedures and handling public money.** (Notice that you don't use a comma right at the end but finish off the list with the word '**and**'.)

In the above example a colon is used to introduce a list ('**includes:** customer care …').

Apostrophes (') are used to show

- possession – for example, **Pete's**
- and omission – for example, '**Don't use those stairs**', rather than 'Do not use those stairs'.

Top tip

A proper sentence contains a subject and a verb.

Spelling strategies

Always use the spell checker facility when compiling anything written, especially when it is a formal piece of writing – this includes email. Never use text talk (LOL).

In Microsoft Word, it is better to select Review from the tool bar and then select spelling and grammar, because this opens out the whole spell checker window. Ordinary spell checkers, however, do not always pick up on homophones and homonyms. 'What are homophones and homonyms?' I hear you ask!

Homonyms are two or more words with the same spelling and pronunciation but with different meanings – for example, flag pole and North Pole.

Homophones are words that are pronounced the same way but are different in meaning and spelling – for example:

Top tip

Make a note of any words that you have difficulty spelling or using in their correct context.

- their and there
- write and right
- by and buy
- witch and which

Most people are well aware of which words they have difficulty with, therefore a good tip is to write these words down, including their meaning and where possible their context; for example:

Over there **is correct**
Over their **is incorrect**

Witch one **is incorrect**
Which one **is correct**

We were away at the time **is correct**
We where away at the time **is incorrect**

Making a personal list will enable you to cross reference the use of these words.

Why not explore basic grammar further through the following link: www. sheppardsoftware.com/grammar/nouns.htm

Equally, the BBC Skillswise website – www.bbc.co.uk/skillswise/0/ – is an excellent place for both literacy and numeracy exercises, including some very interesting games. Why not investigate further and brush up on your skills?

Sample CV

Gwilym Owen

12 Church Terrace

Stateside

Llangefni, Angelsey, LT66 F56

Mob: 099799293666

he@hotmail

Profile

I am a competent and conscientious individual, currently studying a Level 1 Diploma in Electrical Installation at Coleg Menai. This course has supplemented my previous experience of completing a vocational pathway course through school, again with Coleg Menai, whereby I was introduced to engineering principles and health and safety initiatives. I am very ambitious and intend to work hard in gaining an electrical installation apprenticeship.

Key skills

- Basic understanding of engineering principles
- Experience in speaking and dealing with customer relations
- Experienced of carrying out basic health and safety initiatives
- Experienced in undertaking basic stock taking procedures
- Computer literate.

Experience summary

2015   Retail (Saturday Job)

2014–2016   School pathway programme

Training and qualifications

- NVQ Level 1 in engineering

- Certificate in engineering fabrication

- GCSE: English (C), Welsh (E), Science (C), Mathematics (C), ICT (A), Food Technology (F)

Current studies

- Level 1 Diploma in Electrical Installation
- Welsh Baccalaureate
- Essential Skills: Communication, Application of Number, ICT

Hobbies and interests

Rugby, music and surfboarding.

References

On request

Example of a Cover Letter

Your contact information:

Name

Address

Phone Number

Email Address

Date

Employer contact information (if you have it):

Name

Title

Company

Address

Reference if answering an advert

Salutation:

Dear Mr/Ms Last Name, or Dear Sir,

I am writing to you because I am seeking an apprenticeship and I believe that your company has employed young people in the past. I would therefore respectfully like to take the opportunity to express an interest if such a position was to arise in the future.

I believe I possess many of the qualities you will be looking for, because I am hard working, polite and trustworthy and can work well in both a team environment or on my own initiative. I am ambitious and especially eager to learn new skills in the electrical installation industry to further the experience I have already gained during my current work based placement with Mr Pete Roberts, an electrical contractor.

I also have other relevant qualities and experience, such as dealing with customers as well as carrying out basic stock taking, health and safety and hygiene instructions, gained when working in a family run retail outlet. Such experience shows that I can work well under pressure.

I look forward to hearing from you and, with sufficient notice, I can be available for an interview. I enclose a copy of my CV for further information.

Yours faithfully,

Glossary of terms

T	Tera	10^{12}
G	Giga	10^9
M	Mega or meg	10^6
k	Kilo	10^3
d	Deci	10^{-1}
c	Centi	10^{-2}
m	Milli	10^{-3}
μ	Micro	10^{-6}
n	Nano	10^{-9}
p	Pico	10^{-12}

Acceleration
Acceleration is the rate at which an object changes its speed.

Accident
An accident is defined as an unplanned event causing injury or damage to a person or property.

Approved test equipment
Certain test equipment is approved for use by the HSE with regards to carrying out safe isolation procedures as described in GS 38.

Alternating current
Alternating current flows one way and then the other. It is used in the transmission and distribution of electricity because transformers cannot operate with direct current. Domestic mains is therefore an alternating supply.

Ampere
Ampere sometimes shortened to amps, is the SI unit of electrical current. Protective devices such as fuses and circuit breakers are rated in amps.

Basic protection
Basic protection is defined as stopping a person from coming in contact with live supplies. There are various ways of achieving this: insulation, placing live parts out of reach, within enclosures or by installing barriers.

Bonding
Bonding is the linking together of the exposed or extraneous metal parts of an electrical installation such as metal gas and water pipes.

Built environment
All the material that we use in the construction industry.

Cable tray
Cable tray is a type of containment system made from sheet-steel channel that contains multiple holes which aids ventilation. It is mostly used to support SWA cables.

Cable tie
A cable tie is a type of fastener, used to hold bundles of cables together or to a fixed saddle point.

Circuit breaker
Type of circuit protective device that contains both a magnetic trip for very large fault currents and a bi-metallic trip for overloads.

Circuit protective conductor (c.p.c)
A c.p.c connects all exposed conductive parts of an electrical installation to the earthing system. It offers any fault current a low resistance path.

Compact fluorescent lamps (CFLs)
CFLs are energy efficient lamps, which are miniature versions of the larger fluorescent type lamps and were introduced to replace ordinary GLS lamps.

Conductor	A conductor is a material which contains a lot of loosely bound negatively charged electrons which when attracted to a positive energy source cause electrical current to flow.
Conduit	Conduit is a tube, normally made from plastic or metal, that acts as a wiring enclosure, whereby non-sheathed cables are drawn through.
Direct current	Direct current (d.c.) only flows in one direction and has a constant value. Batteries produce d.c.
Earthing	Earthing is a system which connects all exposed conductive parts of an installation to the main protective earthing terminal of the installation. Earthing forms a low resistance path so that a very high fault current will develop, which will enable the circuit protective device to operate very quickly and isolate the circuit.
Electric current	An electric current is caused through the movement of negatively charged electrons. Electric current is measured in amperes but is given the symbol I.
Electric shock	An electric shock occurs when a person becomes part of an electrical circuit.
Environment	The environment describes our natural world, the world that surrounds us.
Emergency alarm call points	Manually operated emergency alarm call points should be provided in all parts of a building, but especially in locations such as workshops and laboratories.
Expansion bolts	Expansion bolts are a type of fixing used in stonework. The most well-known expansion bolt is made by Rawlbolt.
Exposed conductive parts	Any metal element that forms part of an electrical installation is known as an exposed conductive part. Examples include metal conduit, trunking and metal back boxes.
Extraneous conductive parts	Any metal element that does **not** form part of an electrical installation is known as an exposed conductive part. Examples include, structural steelwork, radiators, sinks and gas and water pipes.
Fault protection	Although there are many ways of providing fault protection, the two primary ways involve protective bonding and automatic disconnection of the supply through earthing.
First aid	First aid is defined as the assistance given to any person suffering a sudden illness or injury, with care provided to preserve life and prevent the condition from worsening.
Flexible conduit	Flexible conduit is made from interlinked metal spirals and is largely used to provide the final connection to electrical circuits. The interlinked metal spirals specifically designed to offset any damage that could occur from vibration.
Fluorescent lamp	A fluorescent lamp is a type of lamp that produces light by passing electricity through low-pressure mercury gas.
Fossil fuels	Fossilised remains of prehistoric plants and animals that have produced oil, gas and coal.
Fuse	A fuse is designed to be the weakest part of a circuit. Under fault conditions the excess heat produced will melt its fusing wire.
Green	Processes and technology that are renewable and friendly to the natural environment.
Hazard	A hazard is something with the 'potential' to cause harm.
Health and Safety Executive	Organisation that polices health and safety.
Insulator	An insulator is any material in which its electrons are very firmly bound inside and, therefore, will not allow heat or electricity to flow easily.

Isolation	Isolation is defined as removing all forms of energy from a circuit.
Job sheets	A job sheet or job card lists the actual work required.
Junction box	A type of plastic electrical connections container, used as an alternative way of wiring lighting
Landfill	Materials that cannot be reused or recycled are sent to a landfill site.
Loop-in	Type of wiring used for domestic lighting circuits.
Mass	Mass is a measure of the amount of material in a substance.
Metallic trunking	Metallic trunking is a type of containment system formed from mild steel sheet or plastic.
Mutual inductance	Mutual inductance is the operating principle of a transformer. If electricity is made to flow through its primary coil, then through an electromagnetic link, electricity will be induced in its secondary coil.
Non-statutory regulations	Although not a legal document, they are written in such a way that they meet any legal obligation. For example BS 7671 the Wiring Regulations.
Ohm's law	Ohm's law is used to show the relationship between voltage, current and resistance.
Overload current	An overload occurs when too much is asked of an otherwise healthy circuit.
Plastic plugs	Type of hollow plastic tube fixing used in brick.
Ozone layer	A layer of naturally occurring gas that sits above the surface of the earth.
Polyvinylchloride (PVC)	Type of thermoplastic used in forming cable sheath and inner insulation.
Potential difference	Measured in volts, it is the difference in electrical potential between two points.
Renewables	Resources that are naturally replenished, such as solar energy, rain and wind.
Relay	A relay is an electromagnetic device which incorporates a solenoid and a set of contacts.
Resistance	Resistance is defined as the opposition to current flow.
Short-circuit	A short-circuit is a fault condition, whereby the line conductor comes in to contact with the neutral conductor and bypassing the load.
SI units	SI units are an international system which state how fundamental units of measurement are derived and defined.
Single PVC unsheathed insulated conductors	Also referred to as singles, these type of cables are usually drawn into wiring enclosures such as conduit and trunking. Must not be used outside of these wiring enclosures.
Socket outlets	Socket outlets are used to provide a means of connecting portable electrical appliances to a source of electric supply.
Statutory regulations	Statutory regulations are acts of Parliament and have, therefore, become law.
Thermoplastic	Type of plastic that can be recycled.
Thermosetting	Type of plastic that cannot be recycled. Typical uses include certain plug tops and switch plates.
Time sheets	A time sheet is a form that an employee fills out, which details all their weekly activities.
Transformer	A transformer is an electromagnetic device which is used to step voltage up and down in the transmission and distribution of electricity. Transformers only work on alternating current.
Transferability	Transferable skills are general skills you can use in many other jobs and sectors.

Trunking

Trunking is a plastic or metal type wiring enclosure, usually square or rectangular in shape. Because it contains only one removable side, cables are easily installed in comparison to conduit, which requires cables to be drawn through in stages.

Visual inspection

The visual inspection of an electrical installation is carried out before any testing stage and involves using many of our human senses, such as sight, hearing and touch.

Index

Page numbers in *italics* refer to figures. Page numbers in **bold** refer to tables.